T0180978

Developments in Mathematics

VOLUME 42

More information about this series at http://www.springer.com/series/5834

V. Lakshmibai • Justin Brown

The Grassmannian Variety

Geometric and Representation-Theoretic Aspects

V. Lakshmibai
Department of Mathematics
Northeastern University
Boston, MA, USA

Justin Brown
Department of Mathematics
Olivet Nazarene University
Bourbonnais, IL, USA

ISSN 1389-2177 ISSN 2197-795X (electronic)
Developments in Mathematics
ISBN 978-1-4939-5608-1 ISBN 978-1-4939-3082-1 (eBook)
DOI 10.1007/978-1-4939-3082-1

Mathematics Subject Classification (2010): 14M15, 14L24, 13P10, 14D06, 14M12

Springer New York Heidelberg Dordrecht London

Softcover reprint of the hardcover 1st edition 2015

Printed on acid-free paper

Springer Science+Business Media LLC New York is part of Springer Science+Business Media (www.springer.com)

Preface

This monograph represents an expanded version of a series of lectures given by V. Lakshmibai on Grassmannian varieties at the workshop on "Geometric Representation Theory" held at the Institut Teknologi Bandung in August 2011. While giving the lectures at the workshop, Lakshmibai realized the need for an introductory book on Grassmannian varieties that would serve as a good resource for learning about Grassmannian varieties, especially for graduate students as well as researchers who want to work in this area of algebraic geometry. Hence, the creation of this monograph.

This book provides an introduction to Grassmannian varieties and their Schubert subvarieties, focusing on the treatment of geometric and representation theoretic aspects. After a brief discussion on the basics of commutative algebra, algebraic geometry, cohomology theory, and Gröbner bases, the Grassmannian variety and its Schubert subvarieties are introduced. Following introductory material, the standard monomial theory for the Grassmannian variety and its Schubert subvarieties is presented. In particular, the following topics are discussed in detail:

- the construction of explicit bases for the homogeneous coordinate ring of the Grassmannian and its Schubert varieties (for the Plücker embedding) in terms of certain monomials in the Plücker coordinates (called standard monomials);
- the use of the standard monomial basis,
- the presentation of a proof of the vanishing of the higher cohomology groups of powers of the restriction of the tautological line bundle of the projective space (giving the Plücker embedding).

Further to using the standard monomial basis, the book discusses several geometric consequences, such as Cohen–Macaulayness, normality, unique factoriality, Gorenstein-ness, singular loci, etc., for Schubert varieties, and presents two kinds of degenerations of Schubert varieties, namely, degeneration to a toric variety, and degeneration to a monomial scheme. Additionally, the book presents the relationship between the Grassmannian and classical invariant theory. Included is a discussion on determinantal varieties—their relationship to Schubert varieties as well as to classical invariant theory. The book is concluded with a brief account of some topics related to the flag and Grassmannian varieties: standard monomial theory for a general G/Q, homology and cohomology of the flag variety, free resolutions of Schubert singularities, Bott–Samelson varieties, Frobenius splitting, affine flag varieties, and affine Grassmannian varieties.

This text can be used for an introductory course on Grassmannian varieties. The reader should have some familiarity with commutative algebra and algebraic geometry. A basic reference to commutative algebra is [21] and algebraic geometry [28]. The basic results from commutative algebra and algebraic geometry are summarized in Chapter 2. We have mostly used standard notation and terminology and have tried to keep notation to a minimum. Throughout the book, we have numbered theorems, lemmas, propositions, etc., in order according to their chapter and section; for example, 5.1.3 refers to the third item of the first section in the fifth chapter.

Acknowledgments: V. Lakshmibai thanks CIMPA, ICTP, UNESCO, MESR, MICINN (Indonesia Research School), as well as the organizers, Intan Muchtadi, Alexander Zimmermann of the workshop on "Geometric Representation Theory" held at the Institut Teknologi Bandung, Bandung, Indonesia, August 2011, for inviting her to give lectures on "Grassmannian Variety," and also the Institute for the hospitality extended to her during her stay there.

Both authors thank Reuven Hodges for his feedback on some of the chapters. We thank the makers of ShareLaTeX for making this collaboration easier. We thank the referee for some useful comments, especially, pertaining to Chapter 11.

Finally, J. Brown thanks Owen, Evan, and Callen, for constantly reminding him just how fun life can be.

Boston, MA, USA V. Lakshmibai
Bourbonnais, IL, USA Justin Brown

Contents

Part III Flag Varieties and Related Varieties

Chapter 1
Introduction

This book is an expanded version of a series of lectures given by V. Lakshmibai on Grassmannian varieties at the workshop on "Geometric Representation Theory" held at the Institut Teknologi Bandung, Bandung, Indonesia, in August 2011. In this book, we have attempted to give a complete, comprehensive, and self-contained account of Grassmannian varieties and the Schubert varieties (inside a Grassmannian variety).

In algebraic geometry, Grassmannian varieties form an important fundamental class of projective varieties. In terms of importance, they are second only to projective spaces; in fact, a projective space itself is a certain Grassmannian. A Grassmannian variety sits as a closed subvariety of a certain projective space, the embedding being known as the celebrated Plücker embedding (as described in the next paragraph). Grassmannian varieties are important examples of homogeneous spaces; they are of the form $GL_n(K)/P$, P being a certain closed subgroup (for the Zariski topology) of $GL_n(K)$ (here, $GL_n(K)$ is the group of $n \times n$ invertible matrices with entries in the [base] field K which is supposed to be an algebraically closed field of arbitrary characteristic). In particular, a Grassmannian variety comes equipped with a $GL_n(K)$-action; in turn, the (homogeneous) coordinate ring (for the Plücker embedding) of a Grassmannian variety acquires a $GL_n(K)$-action, thus admitting representation-theoretic techniques for the study of Grassmannian varieties. Thus, Grassmannian varieties are at the crossroads of algebraic geometry, commutative algebra, and representation theory; their study is further enriched by their combinatorics. Schubert varieties in a Grassmannian variety form an important class of subvarieties, and provide a powerful inductive machinery for the study of Grassmannian varieties; in fact, a Grassmannian variety itself is a certain Schubert variety.

A Grassmannian variety (as a set) is the set of all subspaces of a given dimension d in K^n, for some $n \in \mathbb{N}$; it has a canonical projective embedding (the Plücker embedding) via the map which sends a d-dimensional subspace to the associated point in the projective space $\mathbb{P}(\Lambda^d K^n)$. A Grassmannian variety may be thought of

V. Lakshmibai, J. Brown, *The Grassmannian Variety*,
Developments in Mathematics 42, DOI 10.1007/978-1-4939-3082-1_1

as a partial flag variety: given a bunch of r distinct integers $\underline{d} := 1 \leq d_1 < d_2 < \ldots < d_r \leq n-1$, the *partial flag variety* $\mathcal{F}l_{\underline{d}}$, consists of partial flags of type $\underline{d}$, namely sequences $V_{d_1} \subset V_{d_2} \subset \ldots \subset V_{d_r}$, V_{d_i} being a K-vector subspace of K^n of dimension d_i. The extreme case with $r = 1$, corresponds to the *Grassmannian variety* $G_{d,n}$ consisting of d-dimensional subspaces of K^n. If $d = 1$, then $G_{1,n}$ is just the $(n-1)$-dimensional projective space $\mathbb{P}_K^{n-1}$ (consisting of one-dimensional subspaces in K^n). For $r = n-1$, we get the celebrated *flag variety* $\mathcal{F}l_n$, consisting of flags in K^n, where a (full) flag is a sequence $(0) = V_0 \subset V_1 \subset \ldots \subset V_n = K^n$, $\dim V_i = i$. The flag variety $\mathcal{F}l_n$ has a natural identification with the homogeneous space $GL_n(K)/B$, B being the (Borel) subgroup of $GL_n(K)$ consisting of upper triangular matrices.

Throughout the 20th century, mathematicians were interested in the study of the Grassmannian variety and its Schubert subvarieties (as well as the flag variety and its Schubert subvarieties). We shall now mention some of the highlights of the developments in the 20th century on the Grassmannian and the flag varieties, pertaining to the subject matter of this book.

In 1934, *Ehresmann* (cf. [20]) showed that the classes of Schubert subvarieties in the Grassmannian give a $\mathbb{Z}$-basis for the cohomology ring of the Grassmannian, and thus established a key relationship between the geometry of the Grassmannian varieties and the theory of characteristic classes. This result of Ehresmann was generalized by *Chevalley* (cf. [14]) in 1956. Chevalley showed that the classes of the Schubert varieties (in the *generalized flag variety* G/B, G a semisimple algebraic group and B a Borel subgroup) form a $\mathbb{Z}$-basis for the Chow ring of the generalized flag variety. The results of Ehresmann and Chevalley were complemented by the work of *Hodge* (cf. [31, 32]). Hodge developed the *Standard Monomial Theory* for Schubert varieties in the Grassmannian. This theory consists in constructing explicit bases for the homogeneous coordinate ring of the Grassmannian and its Schubert varieties (for the Plücker embedding) in terms of certain monomials (called standard monomials) in the Plücker coordinates. Hodge's theory was generalized to G/B, for G classical by *Lakshmibai, Musili,* and *Seshadri* in the series Geometry of G/P I-V (cf. [49, 50, 53, 56, 82]) during 1975–1986; conjectures were then formulated (cf. [55]) by Lakshmibai and Seshadri in 1991 toward the generalization of Hodge's theory to exceptional groups. These conjectures were proved by *Littelmann* (cf. [65, 67, 68]) in 1994–1998, thus completing the standard monomial theory for semisimple algebraic groups. This theory has led to many interesting and important geometric and representation-theoretic consequences (see [44, 46, 48, 52, 54, 57, 65, 67, 68]).

In this book, we confine ourselves to the Grassmannian varieties and their Schubert subvarieties, since our goal is to introduce the readers to Grassmannian varieties fairly quickly, minimizing the technicalities along the way. We have attempted to give a complete and comprehensive introduction to the Grassmannian variety — its geometric and representation-theoretic aspects.

This book is divided into three parts. Part I is a brief discussion on the basics of commutative algebra, algebraic geometry, cohomology theory, and Gröbner bases. Part II is titled "Grassmann and Schubert Varieties." We introduce the Grassmannian variety and its Schubert subvarieties in Chapter 5. In Chapter 5, we also present the standard monomial theory for the Grassmannian variety and its Schubert subvarieties, namely construction of explicit bases for the homogeneous coordinate ring of the Grassmannian and its Schubert varieties (for the Plücker embedding) in terms of certain monomials in the Plücker coordinates (called standard monomials); we present a proof of the vanishing of the higher cohomology groups of powers of the restriction of the tautological line bundle of the projective space (giving the Plücker embedding), using the standard monomial basis. In Chapter 6, we deduce several geometric consequences—such as Cohen–Macaulayness, normality, a characterization for unique factoriality for Schubert varieties—in fact, these properties are established even for the cones over Schubert varieties (for the Plücker embedding). In addition, we describe the singular locus of a Schubert variety. In Chapter 7, we show that the generators for the ideal of the Grassmannian variety (as well as a Schubert variety) given by the Plücker quadratic relations give a Gröbner basis. We have also presented two kinds of degenerations of Schubert varieties, namely degeneration of a Schubert variety to a toric variety and degeneration to a monomial scheme. We also give a characterization for Gorenstein Schubert varieties.

Part III is titled "Flag Varieties and Related Varieties," and begins with Chapter 8, where we have included a brief introduction to flag varieties and statement of results on the standard monomial theory for flag varieties, as well as degenerations of flag varieties. In Chapter 9, we present the relationship between the Grassmannian and classical invariant theory. In Chapter 10, we present determinantal varieties, their relationship to Schubert varieties as well as to classical invariant theory. In Chapter 11, we give a brief account of some topics related to the flag and Grassmannian varieties: standard monomial theory for a general G/Q, homology and cohomology of the flag variety, free resolutions of Schubert singularities, Bott–Samelson varieties, Frobenius splitting, affine flag varieties, and affine Grassmannian varieties.

Part I
Algebraic Geometry: A Brief Recollection

Chapter 2
Preliminary Material

This chapter is a brief review of commutative algebra and algebraic geometry. We have included basic definitions and results, but omitted many proofs. For details in commutative algebra, we refer the reader to [21, 72] and in algebraic geometry to [28, 75].

2.1 Commutative Algebra

In this section we list the necessary definitions and preliminary results from commutative algebra.

Definition 2.1.1. A ring R is *Noetherian* if every ideal is finitely generated, or equivalently, if every increasing chain of ideals terminates (this is called the *ascending chain condition*).

Theorem 2.1.2 (Hilbert Basis Theorem, cf. Theorem 1.2 of [21]). *If a ring R is Noetherian, then the polynomial ring $R[x_1, \ldots, x_n]$ is Noetherian.*

Note that any field is Noetherian (since a field has no nontrivial, proper ideals).

An ideal P of a ring R is *prime* if $P \neq R$ and $xy \in P$ implies $x \in P$ or $y \in P$. Given an ideal I of R, the *radical* of I is the set $\{r \in R \mid r^n \in I, \text{ for some } n \in \mathbb{N}\}$ and is denoted $\sqrt{I}$. We have $\sqrt{(0)}$ equals the intersection of all prime ideals of R. If $\sqrt{(0)} = (0)$, then R is *reduced*.

We say that $S \subset R$ is a multiplicative set if $0 \notin S$, $1 \in S$, and if $a, b, \in S$ then $ab \in S$. If P is a prime ideal of R, then $R \setminus P$ is a multiplicative set. The *ring of fractions* $S^{-1}R$ is constructed using equivalence classes of pairs $(r, s) \in R \times S$ such that $(r_1, s_1) \sim (r_2, s_2)$ if there exists $s \in S$ such that $s(r_1 s_2 - r_2 s_1) = 0$. One denotes the element (r, s) as the fraction $\frac{r}{s}$. Addition and multiplication in $S^{-1}R$ are defined as

V. Lakshmibai, J. Brown, *The Grassmannian Variety*,
Developments in Mathematics 42, DOI 10.1007/978-1-4939-3082-1_2

$$\frac{r_1}{s_1} + \frac{r_2}{s_2} = \frac{r_1s_2 + r_2s_1}{s_1s_2}, \quad \frac{r_1}{s_1} \cdot \frac{r_2}{s_2} = \frac{r_1r_2}{s_1s_2}.$$

We have the natural ring homomorphism $R \to S^{-1}R$, $r \mapsto \frac{r}{1}$.

Theorem 2.1.3 (cf. Theorem 4.1 of [72]). *The prime ideals of $S^{-1}R$ correspond bijectively to the prime ideals of R disjoint from S.*

Definition 2.1.4. A *zero divisor* of a ring R is a nonzero element $a \in R$ such that there exists a nonzero element $b \in R$ such that $ab = 0$. A nonzero element $a \in R$ that does not satisfy this definition is called a *nonzero divisor*.

When S is the set of all nonzero divisors in R, we call $S^{-1}R$ the *full ring of fractions of R*. If R is an integral domain, then the full ring of fractions is in fact a field, which we call the *field of fractions of R*.

As mentioned above, if P is a prime ideal, then $S = R \setminus P$ is a multiplicative set, and we will denote $S^{-1}R$ as R_P; we call this the *localization* of R with respect to P. Note that R_P is a *local ring*, meaning it has a unique maximal ideal. In this case the maximal ideal is PR_P, the set of all elements in R_P without multiplicative inverses. The prime ideals of R_P correspond to the prime ideals in R contained in P. (Throughout, we use the notation PM when P is an ideal in R and M is an R-module to denote $\{pm \mid p \in P,\ m \in M\}$.) If R is Noetherian, then so is R_P.

For an R-module M, $S^{-1}M$ is defined by using classes of pairs $(m, s) \in M \times S$, where $(m_1, s_1) \sim (m_2, s_2)$ if there exists $s \in S$ such that $s(s_1m_2 - s_2m_1) = 0$. Then $S^{-1}M$ is an $S^{-1}R$-module (in a natural way), and is naturally isomorphic to $S^{-1}R \otimes M$. The functor $M \to S^{-1}M$ from R-modules to $S^{-1}R$-modules is exact, i.e., it takes exact sequences to exact sequences.

Let $R_1 \subset R_2$ be an extension of rings. An element $r \in R_2$ is *integral* over R_1 if r is a zero of some monic polynomial with coefficients in R_1; i.e.,

$$r^n + a_{n-1}r^{n-1} + \ldots + a_1r + a_0 = 0$$

where $a_i \in R_1$. The ring R_2 is *integral over* R_1 if every element of R_2 is integral over R_1. The *integral closure* of R_1 in R_2 is the subring of R_2 consisting of all the elements of R_2 integral over R_1. If R is an integral domain such that R contains all elements of $K(R)$ that are integral over R (here, $K(R)$ is the field of fractions of R, as defined above), then we say R is *integrally closed*.

Definition 2.1.5. If R is a ring such that R_P is an integrally closed domain for every prime ideal $P \subset R$, then R is a *normal ring*.

Example 2.1.6. The ring of integers $\mathbb{Z}$ is a normal ring.

Let $K_1 \subset K_2$ be an extension of fields. The elements $a_1, \ldots, a_d \in K_2$ are *algebraically independent over* K_1 if there does not exist any nonzero polynomial $f(x_1, \ldots, x_d)$ over K_1 such that $f(a_1, \ldots, a_d) = 0$. A maximal subset of algebraically independent elements over K_1 in K_2 is called a *transcendence basis* of K_2/K_1. If K_2

is a finitely generated extension of K_1, then any two transcendence bases of K_2/K_1 have the same cardinality called the *transcendence degree* of K_2 over K_1, and is denoted $\text{tr.deg}_{K_1} K_2$.

Let $R = K[a_1, \ldots, a_n]$ be a finitely generated K-algebra, where R is an integral domain with field of fractions denoted by $K(R)$. Then define $\text{tr.deg}_K R = \text{tr.deg}_K K(R)$.

Definition 2.1.7. For a Noetherian ring R, $\dim R$ (called the Krull dimension of R) is the maximal length t of a strictly increasing chain of prime ideals

$$P_0 \subset P_1 \subset P_2 \subset \ldots \subset P_t \subset R.$$

For example,

1. $\dim \mathbb{Z} = 1$.
2. $\dim K[x_1, \ldots, x_n] = n$, where K is a field.

Definition 2.1.8. If P is a prime ideal of the Noetherian ring R, the *height* of P is defined to be $\dim R_P$, and is denoted $\text{ht}(P)$; the *coheight* of P is defined to be $\dim R/P$.

For the following definitions, let R be a Noetherian ring, and let M be a finitely generated R-module. Our primary examples of R-modules will be ideals of R, localizations of R, or R itself.

Definition 2.1.9. An element $r \in R$ is a *zero divisor* in M if there exists a nonzero $m \in M$ such that $rm = 0$.

Definition 2.1.10. A sequence $r_1, \ldots, r_s \in R$ is an *M-regular sequence* if

1. $(r_1, \ldots, r_i)M \neq M$, and
2. r_i is not a zero divisor in $M/(r_1, \ldots, r_{i-1})M$ for $2 \leq i \leq s$, and r_1 is not a zero divisor in M.

If $r_1, \ldots, r_s$ is an M-regular sequence contained in an ideal I of R, we call it an M-regular sequence in I. If, for any $r_{s+1} \in I$, $r_1, \ldots, r_s, r_{s+1}$ is not an M-regular sequence, we say that $r_1, \ldots, r_s$ is a maximal M-regular sequence in I.

Theorem 2.1.11 (cf. Theorem 16.7 of [72]). *Let R be a Noetherian ring, I an ideal in R, and M a finitely generated R-module. Then any two maximal M-regular sequences in I have the same length.*

Definition 2.1.12. For a local Noetherian ring R with maximal ideal $\mathfrak{m}$ and R-module M, the *depth of M* is defined as the length of a maximal M-regular sequence in $\mathfrak{m}$, and is denoted $\text{depth}\, M$.

Remark 2.1.13. If R is a local Noetherian domain, then $\text{depth}\, R \geq 1$.

Proposition 2.1.14 (cf. Prop. 18.2 of [21]). *Let R be a ring and P a prime ideal, then depth $R_P \leq ht(P)$.*

We remark that depth R_P refers to the depth of R_P as an R_P-module, which is equivalent to what [21] refers to as depth P.

Definition 2.1.15. A ring R such that depth $R_P = \mathrm{ht}(P)$ for every maximal ideal P of R is *Cohen–Macaulay*.

Proposition 2.1.16 (cf. Proposition 18.8 of [21]). *Let R be a ring.*

1. *R is Cohen–Macaulay if and only if R_P is Cohen–Macaulay for every maximal ideal P in R.*
2. *R is Cohen–Macaulay if and only if R_P is Cohen–Macaulay for every prime ideal P in R.*

Proposition 2.1.17 (cf. Theorem. 1.1 of [71] (Graded Case)). *A graded ring R is Cohen–Macaulay if and only if R_P is Cohen–Macaulay for every maximal graded ideal P in R. If in addition, R_0 is a field, then R is Cohen–Macaulay if and only if $R_{\mathfrak{m}}$ is Cohen–Macaulay, $\mathfrak{m}$ being the (unique) maximal graded ideal R_+.*

Lemma 2.1.18 (Nakayama's Lemma, cf. [21]). *Let $(R, \mathfrak{m})$ be a local Noetherian ring, and M a finitely generated R-module. If $\mathfrak{m}M = M$, then $M = (0)$.*

We discuss one consequence of Nakayama's Lemma. Let R, $\mathfrak{m}$, and M be as above, and let $K = R/\mathfrak{m}$. Consider $M/\mathfrak{m}M$ as a vector space over K. Let $x_1, \ldots, x_n \in M$ be such that their classes form a basis in $M/\mathfrak{m}M$ (as a K-vector space). We want to show that $x_1, \ldots, x_n$ generate M. Suppose $N \subseteq M$ is the submodule generated by $x_1, \ldots, x_n$. Then under the canonical map $M \to M/\mathfrak{m}M$, N maps onto $M/\mathfrak{m}M$. Hence $N + \mathfrak{m}M = M$; but $\mathfrak{m} \cdot M/N = (\mathfrak{m}M + N)/N = M/N$. By Nakayama's Lemma, $M/N = (0)$, and $M = N$.

If we now consider $M = \mathfrak{m}$, and find $x_1, \ldots, x_n \in \mathfrak{m}$ that produce a basis for $\mathfrak{m}/\mathfrak{m}^2$, then $x_1, \ldots, x_n$ generates $\mathfrak{m}$, and n is the smallest number of elements in a generating set for $\mathfrak{m}$. We call this number the *embedding dimension* of R, denoted $e(R)$:

$$e(R) = \dim_{R/\mathfrak{m}}(\mathfrak{m}/\mathfrak{m}^2).$$

In general, $\dim R \leq e(R)$.

Definition 2.1.19. $(R, \mathfrak{m})$ is a *regular local ring* if $\dim R = e(R)$. A Noetherian ring R is *regular* if R_P is a regular local ring for every prime ideal $P \subset R$.

Definition 2.1.20. Let R be a ring.

1. An element of R is *prime* if it generates a prime ideal.
2. An element $r \in R$ is *irreducible* if it is not a unit, and whenever $r = st$, either s or t is a unit.

Remark 2.1.21. If $r \in R$ is a prime element, then r is irreducible.

Definition 2.1.22. Let R be an integral domain, R is *factorial* if every nonzero element can be expressed uniquely as a product of irreducible elements, with uniqueness up to factors that are units. We will also call such a ring an *unique factorization domain*, or *UFD*.

Remark 2.1.23. A ring is factorial if and only if every irreducible element is prime (cf. [21, p. 14]).

Remark 2.1.24. We conclude this section with several facts about regular rings, the proofs of which require too much machinery to include in this text; but we attach references for the interested readers.

1. If R is a regular local ring, and P is a prime ideal, then R_P is a regular local ring, [72, Theorem 19.3].
2. A regular ring is normal, [72, Theorem 19.4].
3. A regular local ring is Cohen–Macaulay, [72, Theorem 17.8].
4. A regular local ring is factorial, [21, Theorem 19.19].

2.2 Affine Varieties

Let K be the base field, which we suppose to be algebraically closed of arbitrary characteristic. Our primary reference for the following sections is [28].

We shall denote by $\mathbb{A}^n_K$ or just $\mathbb{A}^n$, the affine n-space, consisting of $(a_1, \ldots, a_n)$, $a_i \in K$. For $P = (a_1, \ldots, a_n) \in \mathbb{A}^n$, the a_i's are called the *affine coordinates* of P.

Given an ideal I in the polynomial algebra $K[x_1, \ldots, x_n]$, let

$$V(I) = \{(a_1, \ldots, a_n) \in \mathbb{A}^n \mid f(a_1, \ldots, a_n) = 0 \text{ for all } f \in I\}.$$

The set $V(I)$ is called an *affine variety*. Clearly $V(I) = V(\sqrt{I})$. Fixing a (finite) set of generators $\{f_1, \ldots, f_r\}$ for I, $V(I)$ can be thought of as the set of common zeros of $f_1, \ldots, f_r$.

Conversely, given a subset $X \subset \mathbb{A}^n$, let

$$\mathcal{I}(X) = \{f \in K[x_1, \ldots, x_n] \mid f(x) = 0, \text{ for all } x \in X\}.$$

2.2.1 *Zariski topology on* $\mathbb{A}^n$

Define a topology on $\mathbb{A}^n$ by declaring $\{V(I) \mid I \text{ an ideal in } K[x_1, \ldots, x_n]\}$ as the set of closed sets. We now check that this defines a topology on $\mathbb{A}^n$. We have

1. $\mathbb{A}^n = V((0))$, $\emptyset = V(K[x_1, \ldots, x_n])$.
2. $V(I) \cup V(J) = V(I \cap J)$.
3. $\bigcap_\alpha V(I_\alpha) = V\left(\sum_\alpha I_\alpha\right)$.

Remark 2.2.1.

1. $\mathbb{A}^n$ is a T_1-space for the Zariski topology, i.e., points are closed subsets.
2. $\mathbb{A}^n$ is *not* Hausdorff for the Zariski topology. For instance, consider $\mathbb{A}^1$, the closed subsets are precisely the finite sets, and hence no two nonempty open sets can be disjoint.
3. Closed sets in the Zariski topology satisfy the descending chain condition (i.e., every decreasing chain of closed sets terminates); this follows from the fact that $K[x_1, \ldots, x_n]$ is Noetherian, and ideals in $K[x_1, \ldots, x_n]$ satisfy the ascending chain condition. Hence, open sets satisfy the ascending chain condition; in particular, any nonempty collection of open sets has a maximal element. Hence $\mathbb{A}^n$ is quasicompact, i.e., every open cover admits a finite subcover (the term "quasi" is used since $\mathbb{A}^n$ is not Hausdorff).
4. If $K = \mathbb{C}$, then the zero-set of a polynomial $f \in \mathbb{C}[x_1, \ldots, x_n]$ is closed in the usual topology of $\mathbb{C}^n$, being the inverse image of the closed set $\{0\}$ in $\mathbb{C}$ under the continuous map $\mathbb{C}^n \to \mathbb{C}$, $a \mapsto f(a)$. The set of common zeros of a collection of polynomials is also closed in the usual topology, being the intersection of closed sets. Of course, the complex n-space $\mathbb{C}^n$ has plenty of other closed sets which are not obtained this way (as is clear in the case $n = 1$). Thus the usual topology is stronger than the Zariski topology.
5. We have

$$X \subset V(\mathcal{I}(X)), \; I \subset \mathcal{I}(V(I)).$$

6. In fact, $V(\mathcal{I}(X)) = \bar{X}$, the closure of X.

Theorem 2.2.2 (Hilbert's Nullstellensatz, cf. [75], Chap. 1, Thm. 1). *Let I be an ideal in $K[x_1, \ldots, x_n]$. Then $\sqrt{I} = \mathcal{I}(V(I))$.*

As a consequence of the fact (6) above and Hilbert's Nullstellensatz we obtain an inclusion-reversing bijection

$$\begin{aligned} \{\text{radical ideals in } K[x_1, \ldots, x_n]\} &\longleftrightarrow \{\text{affine varieties in } \mathbb{A}^n\}, \\ I &\longmapsto V(I) \\ \mathcal{I}(X) &\longleftarrow\!\shortmid\; X. \end{aligned}$$

Under the above bijection, points of $\mathbb{A}^n$ correspond one-to-one to the maximal ideals of $K[x_1, \ldots, x_n]$. (The maximal ideal corresponding to $(a_1, \ldots, a_n)$ is the ideal $(x_1 - a_1, \ldots, x_n - a_n)$.)

A topological space X is said to be *irreducible* if X is nonempty and cannot be written as the union of two proper nonempty closed sets in X, or equivalently, any two nonempty open sets in X have a nonempty intersection, or equivalently, any nonempty open set is dense. It is easily seen that a subspace $Y \subset X$ is irreducible if and only if its closure $\bar{Y}$ is irreducible.

A topological space is said to be *Noetherian* if every open set in X is quasicompact, or equivalently, if open sets satisfy the maximal condition (i.e., a nonempty collection of open sets has a maximal element), or equivalently, if open sets satisfy the ascending chain condition, or equivalently, if closed sets satisfy the descending chain condition. As an example, the Noetherian property of $K[x_1, \ldots, x_n]$ implies the descending chain condition on the set of affine varieties in $\mathbb{A}^n$.

Let X be a Noetherian topological space. It can be seen easily that every nonempty closed subset is a finite union of irreducible closed subsets. In particular X is a finite union of (maximal) irreducible closed subsets, and these are called the irreducible components of X.

Proposition 2.2.3 (cf. Chap. I, Cor. 1.4 of [28]). *A closed set X in $\mathbb{A}^n$ is irreducible if and only if $\mathcal{I}(X)$ is prime. In particular, $\mathbb{A}^n$ is irreducible.*

Corollary 2.2.4. *Under the bijection following Theorem 2.2.2, the prime ideals correspond to irreducible affine varieties.*

For an affine variety X, we take $\mathcal{I}(X)$ to be a radical ideal, in view of the bijection following Theorem 2.2.2.

2.2.2 The affine algebra $K[X]$

A finitely generated K-algebra is also called an *affine K-algebra* (or simply an affine algebra, the field K being fixed). Let X be an affine variety in $\mathbb{A}^n$. The affine algebra $K[x_1, \ldots, x_n]/\mathcal{I}(X)$ is called the *affine algebra of* X and is denoted $K[X]$. Now each $f \in K[X]$ defines a function $X \to K$, $a \mapsto f(a)$ (note that this is well defined). Thus each element of $K[X]$ may be thought of as a polynomial function on X (with values in K). For this reason, $K[X]$ is also called the *algebra of polynomial functions on* X, or also the *algebra of regular functions on* X, or the *coordinate ring of* X. If X is irreducible, then $K[X]$ is an integral domain (since $\mathcal{I}(X)$ is prime), and the quotient field $K(X)$ of $K[X]$ is called the *function field of* X (or the *field of rational functions on* X).

As in the case of $\mathbb{A}^n$, we see that we have a bijection between closed subsets of X and the radical ideals of $K[X]$, under which the irreducible closed subsets of X correspond to the prime ideals of $K[X]$. In particular, the points of X are in one-to-one correspondence with the maximal ideals of $K[X]$. Further, X is a Noetherian topological space, and the *principal open subsets* $X_f = \{x \in X \mid f(x) \neq 0\}, f \in K[X]$, give a base for the Zariski topology.

Let $X \subset \mathbb{A}^n$, $Y \subset \mathbb{A}^m$ be two affine varieties. A *morphism* $\varphi : X \to Y$ is a mapping of the form $\varphi(a) = (\psi_1(a), \ldots, \psi_m(a))$, where $a \in X$, and for $i = 1, \ldots, m$, $\psi_i \in K[x_1, \ldots, x_n]$. A morphism $\varphi : X \to Y$ defines a K-algebra morphism $\varphi^* : K[Y] \to K[X]$ given by $\varphi^*(f) = f \circ \varphi$. We have

Theorem 2.2.5 (cf. Chap. I, Cor. 3.8 of [28]). *The map $X \mapsto K[X]$ defines a (contravariant) equivalence of the category of affine varieties (with morphisms as defined above) and the category of affine K-algebras (i.e., finitely generated K-algebras) without nonzero nilpotents (with K-algebra maps as morphisms). Further, irreducible affine varieties correspond to affine K-algebras which are integral domains.*

2.2.3 Products of affine varieties

The product of the Zariski topologies on $\mathbb{A}^n$ and $\mathbb{A}^m$ does not give the Zariski topology on $\mathbb{A}^{n+m}$. For example, in $\mathbb{A}^1 \times \mathbb{A}^1$ the only closed sets in the product topology are finite unions of horizontal and vertical lines, while $\mathbb{A}^2$ has many more sets that are closed in the Zariski topology.

To arrive at a correct definition (so that we will have $\mathbb{A}^n \times \mathbb{A}^m \cong \mathbb{A}^{n+m}$), one takes the general category-theoretical definition, and defines the product as a triple (Z, p, q), where Z is an affine variety and $p : Z \to X$, $q : Z \to Y$ are morphisms such that given a triple (M, α, β), where M is an affine variety and $\alpha : M \to X$, $\beta : M \to Y$ are morphisms, there exists a unique morphism $\theta : M \to Z$ such that the following diagram is commutative:

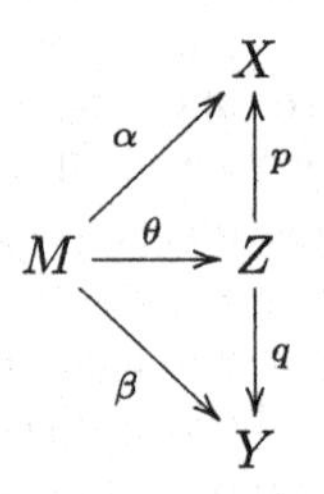

Theorem 2.2.6 (Existence of products). *Let X, Y be two affine varieties with coordinate rings R, S, respectively. Then the affine variety Z with coordinate ring $R \otimes_K S$ together with the canonical maps $p : Z \to X$, $q : Z \to Y$ (induced by $R \to R \otimes S$, $r \mapsto r \otimes 1$, and $S \to R \otimes S$, $s \mapsto 1 \otimes s$, respectively) is a product of X and Y.*

The uniqueness up to isomorphism of a product follows from the universal mapping property of a product. The product of X and Y is denoted by $(X \times Y, p, q)$.

Remark 2.2.7. Let $X \subset \mathbb{A}^n$, $Y \subset \mathbb{A}^m$ be two affine varieties. Then the product variety $X \times Y$ (as defined above) is nothing but the set $X \times Y \subset \mathbb{A}^{n+m}$, together with the induced topology.

2.3 Projective Varieties

We begin this section by introducing the projective space $\mathbb{P}^n$. We shall denote by $\mathbb{P}^n_K$, or just $\mathbb{P}^n$, the set $\left(\mathbb{A}^{n+1} \setminus \{0\}\right) / \sim$, where $\sim$ is the equivalence relation $(a_0, \ldots, a_n) \sim (b_0, \ldots, b_n)$ if there exists $\lambda \in K^*$ such that $(a_0, \ldots, a_n) = \lambda (b_0, \ldots, b_n)$. Thus a point $P \in \mathbb{P}^n$ is determined by an equivalence class $[a_0, \ldots, a_n]$, and for any $(n+1)$-tuple $(b_0, \ldots, b_n)$ in this equivalence class, the b_i's will be referred as the *projective* (or *homogeneous*) *coordinates of* P.

Sometimes we write $\mathbb{P}^n$ also as $\mathbb{P}(V)$, where V is an $(n+1)$-dimensional K-vector space, the points of $\mathbb{P}^n$ being identified with one-dimensional subspaces of V. Let $f(x_0, \ldots, x_n) \in K[x_0, \ldots, x_n]$. Further, let f be homogeneous of degree d. The homogeneity of f implies $f(\lambda x_0, \ldots, \lambda x_n) = \lambda^d f(x_0, \ldots, x_n)$, $\lambda \in K^*$. Hence it makes sense to talk about f being zero or nonzero at a point $P \in \mathbb{P}^n$.

Let I be a homogeneous ideal in $K[x_0, \ldots, x_n]$, i.e., for each $f \in I$, the homogeneous parts of f belong to I, or equivalently, I is generated by some set of homogeneous polynomials. Let

$$V(I) = \{P \in \mathbb{P}^n \mid f(P) = 0, \text{ for all homogeneous } f \in I\}.$$

The set $V(I)$ is called a *projective variety*. Conversely, given a subset of $X \subset \mathbb{P}^n$, let $\mathcal{I}(X)$ be the ideal generated by

$$\{f \in K[x_0, \ldots, x_n], f \text{ homogeneous} \mid f(P) = 0 \text{ for all } P \in X\}.$$

As in the affine case, $\mathcal{I}(X)$ is a radical ideal. We have a similar version (as in the affine case) of the Nullstellensatz with one minor adjustment, namely the ideal I_0 in $K[x_0, \ldots, x_n]$ generated by $x_0, \ldots, x_n$ is a proper radical ideal, but clearly has no zeros in $\mathbb{P}^n$. Deleting I_0, we have a similar formulation given by the following.

Theorem 2.3.1. *The maps $I \mapsto V(I)$, $X \mapsto \mathcal{I}(X)$ define an inclusion-reversing bijection between the set of homogeneous radical ideals of $K[x_0, \ldots, x_n]$ other than I_0 and the projective varieties in $\mathbb{P}^n$.*

2.3.1 Zariski topology on $\mathbb{P}^n$

The Zariski topology on $\mathbb{P}^n$ is defined in exactly the same way as in the affine case, by declaring

$$\{V(I) \mid I \text{ a homogeneous ideal in } K[x_0, \ldots, x_n] \text{ other than } I_0\}$$

as closed sets. As in the affine case, under the above bijection, the homogeneous prime ideals (other than I_0) correspond to irreducible projective varieties. Let $U_i = \{[a] \in \mathbb{P}^n \mid a_i \neq 0\}$, $0 \leq i \leq n$. (These are some special open sets.) The map $U_i \to \mathbb{A}^n$,

$$[a] \mapsto \left(\frac{a_0}{a_i}, \dots, \frac{a_{i-1}}{a_i}, \frac{a_{i+1}}{a_i}, \dots, \frac{a_n}{a_i} \right)$$

defines an isomorphism of affine varieties. The quotients

$$\frac{a_0}{a_i}, \dots, \frac{a_{i-1}}{a_i}, \frac{a_{i+1}}{a_i}, \dots, \frac{a_n}{a_i}$$

are called the *affine coordinates on* U_i, $0 \leq i \leq n$. Note that $\{U_i, 0 \leq i \leq n\}$ gives an open cover for $\mathbb{P}^n$.

2.4 Schemes — Affine and Projective

2.4.1 Presheaves

Let X be a topological space. Let top(X) be the category whose objects are open sets in X, and whose morphisms are inclusions. Let $\mathcal{C}$ be a category. A *$\mathcal{C}$-valued presheaf on X* is a contravariant functor $U \mapsto F(U)$ from top(X) to $\mathcal{C}$. Thus, if $V \subset U$ are open sets in X, then we have a $\mathcal{C}$-morphism

$$\mathrm{res}_V^U : F(U) \longrightarrow F(V)$$

called the *restriction map*.

A *morphism of presheaves* $\varphi : F \to F'$ is a morphism of functors. Suppose $\mathcal{C}$ is a category of "sets with structure," like groups, rings, modules, etc. Then we say that F is a presheaf of groups, rings, modules, etc., respectively.

If $x \in X$, then the collection $\mathcal{U}_x = \{F(U), U \text{ open neighborhood of } x\}$ is a directed system, and $F_x = \varinjlim_{F(U) \in \mathcal{U}_x} F(U)$ is called the *stalk of F at x*.

2.4.2 Sheaves

Let F be a $\mathcal{C}$-valued presheaf on X, where $\mathcal{C}$ is some category of "sets with structure." Then F is called a *sheaf* if it satisfies the following "sheaf axioms": For every collection $\{U_i\}$ of open sets in X with $U = \bigcup U_i$,

1. If $f, g \in F(U)$ are such that $f|_{U_i} = g|_{U_i}$ for all i, then $f = g$.
2. If $\{f_i \in F(U_i)\}$ is a collection such that $f_i|_{U_i \cap U_j} = f_j|_{U_i \cap U_j}$ for all i and j, then there exists an $f \in F(U)$ such that $f|_{U_i} = f_i$.

Example 2.4.1. Let X be a topological space, and for U open in X, let $F(U)$ be the ring of continuous real valued functions on U. The assignment $U \mapsto F(U)$, for U open, defines a sheaf.

Example 2.4.2. Let $X \in \mathbb{A}^n$ be an irreducible affine variety, with function field $K(X)$. Let $R = K[X]$. Define

$$\mathcal{O}_{X,x} = \{f \in K(X) \mid f \text{ is regular at } x\},$$

(note that f is regular at x if $f = g/h$, with $g, h \in R,\ h(x) \neq 0$). We have, $\mathcal{O}_{X,x}$ is simply $R_{\mathcal{P}}$, where $\mathcal{P}$ is the prime ideal $\{f \in R \mid f(x) = 0\}$; in particular, $\mathcal{O}_{X,x}$ is a local ring. The assignment $U \mapsto \bigcap_{x \in U} \mathcal{O}_{X,x}$, for U open defines a sheaf called the *structure sheaf*, and denoted $\mathcal{O}_X$. Note that $\mathcal{O}_X(X) = K[X]$.

2.4.3 *Sheafification*

Let F be a $\mathcal{C}$-valued presheaf on X, where $\mathcal{C}$ is some category of "sets with structure." Then there is a sheaf F', called the *sheafification of* F, or the *sheaf associated with* F, and a morphism $f : F \to F'$ such that the map $\mathrm{Mor}(F', G) \to \mathrm{Mor}(F, G)$ (induced by f) is bijective whenever G is a sheaf. Moreover, such an F' is unique.

We now give a construction of F'. Let $E = \bigcup_{x \in X} F_x$ (a disjoint union of sets). Let $p : E \to X$ be the "projection map," namely $p(a) = x$, if $a \in F_x$. For U open in X, and $\sigma \in F(U)$, we have a canonical map (also denoted by σ) $\sigma : U \to E, \sigma(x) = \overline{\sigma_x}$, where $\overline{\sigma_x}$ is the image of σ under the canonical map $F(U) \to F_x$. Equip E with the strongest topology which makes $\sigma : U \to E$ continuous for all $\sigma \in F(U)$, and all open U. It can be seen easily that a set G in E is open if and only if for every open U in X and $\sigma \in F(U)$, the set $W = \{x \in U \mid \overline{\sigma_x} \in G\}$ is open in X. The space (E, p) is called the *etale space of* F.

Let F' be the sheaf of continuous sections of p, i.e., for U open in X,

$$F'(U) = \{s : U \to E \text{ continuous such that } p \circ s = Id_U\}.$$

Remark 2.4.3. Given a sheaf $\mathcal{F}$ on X, $\mathcal{F}(X)$ is usually denoted by $\Gamma(X, \mathcal{F})$, and its elements are called *global sections* of $\mathcal{F}$.

2.4.4 *Ringed and geometric spaces*

A *ringed space* is a topological space X together with a sheaf $\mathcal{O}_X$ of rings (commutative with identity element) on X. A *geometric space* is a ringed space $(X, \mathcal{O}_X)$ whose stalks $\mathcal{O}_{X,x}$ are local rings. The sheaf $\mathcal{O}_X$ is called the structure sheaf of X. We denote the maximal ideal of $\mathcal{O}_{X,x}$ by $\mathfrak{m}_x$, and the residue class field by $K(x)$.

A morphism $(X, \mathcal{O}_X) \to (Y, \mathcal{O}_Y)$ of ringed spaces consists of a continuous map $f : X \to Y$ together with ring homomorphisms

$$f_U^V : \mathcal{O}_Y(V) \to \mathcal{O}_X(U)$$

for $U \subset X$, $V \subset Y$ open sets such that $f(U) \subset V$. These maps are required to be compatible with the respective restriction maps in $\mathcal{O}_X$ and $\mathcal{O}_Y$. Then one can see that if $x \in X$ and $y = f(x)$, then f induces a ring-map at the level of stalks $f_x : \mathcal{O}_{Y,y} \to \mathcal{O}_{X,x}$. A morphism of geometric spaces is a morphism of ringed spaces such that f_x is a local homomorphism, i.e., $f_x(\mathfrak{m}_y) \subset \mathfrak{m}_x$.

2.5 The Scheme Spec(A)

Let A be a commutative ring with identity element, and let Spec(A) be the set of all prime ideals in A. Define a topology on Spec(A) (called the *Zariski topology on* Spec(A)) by declaring the closed sets as $V(I) := \{\mathfrak{p} \in \text{Spec}(A) \mid \mathfrak{p} \supset I\}$, for I any ideal of A. For $Y \subset X = \text{Spec}(A)$, let $\mathcal{I}(Y) = \bigcap_{\mathfrak{p} \in Y} \mathfrak{p}$. Then $V(\mathcal{I}(Y)) = \overline{Y}$. Further, we have $\mathcal{I}(V(I)) = \sqrt{I}$. Thus we have an inclusion-reversing bijection between the set of closed sets in Spec(A), and the set of radical ideals in A, under which irreducible closed sets correspond to prime ideals. If A is Noetherian, then Spec(A) is a Noetherian topological space, and the irreducible components of X correspond to the minimal primes in A.

Let $f \in A$, and $\mathfrak{p} \in \text{Spec}(A)$. Let $f(\mathfrak{p})$ be the image of f in the residue class field of $A_{\mathfrak{p}}$ (which is simply the field of fractions of $A/\mathfrak{p}$). Let $X = \text{Spec}(A)$, and $X_f = X \setminus V((f)) = \{\mathfrak{p} \in X \mid f(\mathfrak{p}) \neq 0\}$. Note that X_f is open, and called a *principal open set*. For any ideal I, we have $V(I) = \bigcap_{f \in I} V((f))$. Thus the principal open sets form a base of the Zariski topology on X.

We will now define a geometric space structure on Spec(A). Let $X = \text{Spec}(A)$, and let $U \subseteq X$ be an open set. Define $\mathcal{O}_X(U)$ to be the set of functions $f : U \to \coprod_{P \in U} A_P$ (where A_P is the localization of A at the prime ideal P) where $f(P) \in A_P$ and f is locally a quotient of elements of A; i.e., for each $P \in U$, there exists an open $V \subseteq U$, $P \in V$, with elements $a, b \in A$ such that for each $P' \in V$, $f(P') = a/b \in A_{P'}$ (where $b \notin P'$). It is easily seen that the stalk $\mathcal{O}_{X,x}$ is simply A_x. Thus we have the following.

Theorem 2.5.1 (cf. Chap. II, Prop. 2.3 of [28]). *$(Spec(A), \mathcal{O}_{Spec(A)})$ is a geometric space.*

In [28], the phrase "locally ringed space" is used in place of geometric space.

Remark 2.5.2. If A is an integral domain with quotient field K, then the A_x's are subrings of K, and $\mathcal{O}_X$ can be defined directly by $\mathcal{O}_X(U) = \bigcap_{x \in U} A_x$.

Let $X = \mathrm{Spec}(A)$, $Y = \mathrm{Spec}(B)$. Then it is seen easily that a morphism $\varphi : (X, \mathcal{O}_X) \to (Y, \mathcal{O}_Y)$ induces a ring homomorphism $B \to A$. Conversely, a ring homomorphism $B \to A$ induces a morphism $X \to Y$ (cf. [28, Chap. II, Prop. 2.3]).

Definition 2.5.3. An *affine scheme* is a geometric space $(X, \mathcal{O}_X)$ which is isomorphic to $(\mathrm{Spec}(A), \mathcal{O}_{\mathrm{Spec}(A)})$ for some ring A.

Remark 2.5.4. The map $A \mapsto (\mathrm{Spec}(A), \mathcal{O}_{\mathrm{Spec}(A)})$ defines a (contravariant) equivalence of the category of commutative rings and the category of affine schemes.

Definition 2.5.5. A *prescheme* is a geometric space $(X, \mathcal{O}_X)$ which has a finite cover by open sets U such that $(U, \mathcal{O}|_U)$ is an affine scheme.

Definition 2.5.6. A prescheme X is called a *scheme* if the diagonal $\Delta(X)$ $(= \{(x,x) \in X \times X\})$ is closed in $X \times X$.

Remark 2.5.7. An affine scheme is a scheme in the sense of Definition 2.5.6.

Definition 2.5.8. A morphism of schemes $\phi : X \to Y$ is called an *immersion* if ϕ is an isomorphism of X with an open subscheme of a closed subscheme of Y.

Definition 2.5.9. A morphism of schemes $\phi : X \to Y$ is called an *open immersion* (resp.*closed immersion*) if ϕ is an isomorphism of X with an open (resp. closed) subscheme Y.

2.6 The Scheme Proj(S)

Let $S = \bigoplus_{d \geq 0} S_d$ be a *graded ring*, i.e., S_d is an abelian group and $S_d S_e \subset S_{d+e}$. Let $S_+ = \bigoplus_{d>0} S_d$. Define Proj$(S)$ to be the set of all homogeneous prime ideals of S not containing S_+. For a homogeneous ideal $\mathfrak{a}$ of S, set

$$V(\mathfrak{a}) = \{\mathfrak{p} \in \mathrm{Proj}(S) \mid \mathfrak{p} \supseteq \mathfrak{a}\}.$$

In view of the following lemma, declaring the set of $V(\mathfrak{a})$, $\mathfrak{a}$,a homogeneous ideal in S as the closed sets, we obtain the Zariski topology on Proj(S).

Lemma 2.6.1 (cf. Chap. II, Lemma 2.4 of [28]).

1. *If $\mathfrak{a}$ and $\mathfrak{b}$ are two homogeneous ideals in S, then $V(\mathfrak{a}\mathfrak{b}) = V(\mathfrak{a}) \cup V(\mathfrak{b})$.*
2. *If $\{\mathfrak{a}_i\}$ is any family of homogeneous ideals in S, then $V(\sum \mathfrak{a}_i) = \bigcap V(\mathfrak{a}_i)$.*

As we did for Spec(A), we define the structure sheaf $\mathcal{O}_X$ on $X = \mathrm{Proj}(S)$. For $\mathfrak{p} \in \mathrm{Proj}(S)$, let $S_{(\mathfrak{p})}$ denote the *homogeneous localization* of S at $\mathfrak{p}$ consisting of elements of degree 0 in $S_{\mathfrak{p}}$, i.e.,

$$S_{(\mathfrak{p})} = \left\{ \frac{f}{g} \in S_{\mathfrak{p}} \mid f, g \text{ homogeneous of the same degree} \right\}.$$

For an open set $U \subseteq \mathrm{Proj}(S)$, define $\mathcal{O}_X(U)$ as the set of functions $s : U \to \coprod S_{(\mathfrak{p})}$ such that the following hold:

1. For each $\mathfrak{p} \in U$, $s(\mathfrak{p}) \in S_{(\mathfrak{p})}$.
2. s is locally a quotient of elements of S, i.e., for each $\mathfrak{p}$ in U, there exists a neighborhood V of $\mathfrak{p}$ in U, and homogeneous elements $g,f \in S$ of the same degree, such that for all $\mathfrak{q} \in V, f \notin \mathfrak{q}$, $s(\mathfrak{q}) = \frac{g}{f}$ in $S_{(\mathfrak{q})}$.

It is clear that $\mathcal{O}_X$ is a presheaf of rings, with the natural restrictions, and it is also clear from the local nature of the definition that $\mathcal{O}_X$ is in fact a sheaf.

Definition 2.6.2. We define $(\mathrm{Proj}(S), \mathcal{O}_{\mathrm{Proj}(S)})$ to be $\mathrm{Proj}(S)$ with the sheaf of rings constructed above.

Proposition 2.6.3 (cf. Chap. II, Prop. 2.5 of [28]). *$(Proj(S), \mathcal{O}_{Proj(S)})$ is a scheme. Further, for $\mathfrak{p} \in Proj(S)$, the stalk $\mathcal{O}_{Proj(S),\mathfrak{p}}$ is isomorphic to the local ring $S_{(\mathfrak{p})}$.*

For a homogeneous element $f \in S_+$, define $D_+(f) = \{\mathfrak{p} \in \mathrm{Proj}(S) \mid f \notin \mathfrak{p}\}$. Then $D_+(f)$ is open in $\mathrm{Proj}(S)$. Further, these open sets cover $\mathrm{Proj}(S)$, and we have an isomorphism of geometric spaces:

$$\left(D_+(f), \mathcal{O}|_{D_+(f)}\right) \cong \mathrm{Spec}(S_{(f)}),$$

where $S_{(f)}$ is the subring of elements of degree 0 in S_f. See [28, Chap. II, §2] for details.

2.6.1 *The cone over X*

Let $X = \mathrm{Proj}(S)$ be a projective variety. Then the affine variety $\hat{X} = \mathrm{Spec}(S)$ is called the *cone over X*.

2.7 Sheaves of $\mathcal{O}_X$-Modules

Definition 2.7.1. Let $(X, \mathcal{O}_X)$ be a scheme. A sheaf F on X is said to be a *sheaf of $\mathcal{O}_X$-modules* if for $U \subset X$ open, $F(U)$ is an $\mathcal{O}_X(U)$-module.

Example 2.7.2. Let M be an A-module. Then M defines a sheaf $\tilde{M}$, namely for all $f \in A$, $\tilde{M}(X_f) = M_f$ (note that for any open U, we have $\tilde{M}(U) = \varprojlim_{X_f \subset U} M_f$). Then $\tilde{M}$ is a sheaf of $\mathcal{O}_X$-modules.

Definition 2.7.3. Let X be a topological space, and R be a ring. We give R the discrete topology (i.e., all subsets of R are open). For any open set $U \subseteq X$, let $F(U)$ be the ring of all continuous maps from U to R. Along with the natural restriction maps, F is a sheaf, called the *constant sheaf* on X, so named because as long as U is open and connected, $F(U) = R$.

Let S be a graded ring, and M a graded S-module, i.e., M is an S-module together with a decomposition $M = \bigoplus_{d \in \mathbb{Z}} M_d$ such that $S_d \cdot M_r \subseteq M_{d+r}$. The sheaf $\tilde{M}$ on Proj(S) is defined as follows. For $\mathfrak{p} \in \text{Proj}(S)$, let $M_{(\mathfrak{p})}$ denote the group of elements of degree 0 in $M_{\mathfrak{p}}$. For an open set $U \subseteq \text{Proj}(S)$, define $\tilde{M}(U)$ as the set of functions $s : U \to \coprod M_{(\mathfrak{p})}$ such that the following holds:

1. For each $\mathfrak{p} \in U$, $s(\mathfrak{p}) \in M_{(\mathfrak{p})}$.
2. s is locally a quotient, i.e., for each $\mathfrak{p}$ in U, there exists a neighborhood V of $\mathfrak{p}$ in U, and homogeneous elements $m \in M, f \in S$ of the same degree, such that for all $\mathfrak{q} \in V, f \notin \mathfrak{q}$, $s(\mathfrak{q}) = \frac{m}{f}$ in $M_{\mathfrak{q}}$.

We make $\tilde{M}$ into a sheaf with the natural restriction maps. We have the following facts:

1. For $\mathfrak{p} \in \text{Proj}(S)$, the stalk $\tilde{M}_{\mathfrak{p}} \cong M_{(\mathfrak{p})}$.
2. For a homogeneous element $f \in S_+$, we have, $\tilde{M}|_{D_+(f)} \cong \left(\widetilde{M_{(f)}}\right)$ via the isomorphism $D_+(f) \cong \text{Spec}(S_{(f)})$, where $M_{(f)}$ denotes the group of elements of degree 0 in the localized module M_f.

See [28, Chap. II, §5] for details.

2.7.1 *The twisting sheaf* $\mathcal{O}_X(1)$

Let $X = \text{Proj}(S)$. For $n \in \mathbb{Z}$, set $M(n)$ to be the graded S-module with $M(n)_d = M_{n+d}$ for all $d \in \mathbb{Z}$. Define the sheaf $\mathcal{O}_X(n)$ to be $\widetilde{S(n)}$. The sheaf $\mathcal{O}_X(1)$ is called the *twisting sheaf of Serre*. For any sheaf F of $\mathcal{O}_X$-modules, we define

$$F(n) = F \otimes_{\mathcal{O}_X} \mathcal{O}_X(n).$$

Let $f : X \to Y$ be a morphism between two schemes. Let $\mathcal{F}$ be any sheaf on X. We define the direct image sheaf $f_*\mathcal{F}$ on Y by $(f_*\mathcal{F})(V) = \mathcal{F}(f^{-1}(V))$ for any open subset V of Y.

Let $f : X \to Y$ be as above. Let $\mathcal{G}$ be any sheaf on Y. We define the sheaf $f^{-1}\mathcal{G}$ on X to be the sheaf associated to the presheaf $U \mapsto \varinjlim_{V \supseteq f(U)} \mathcal{G}(V)$, for U open in X, where the limit is taken over all open subsets V of Y containing $f(U)$.

We define the sheaf $f^*\mathcal{G}$, the inverse image of $\mathcal{G}$ by the morphism f, to be $f^{-1}\mathcal{G} \otimes_{f^{-1}\mathcal{O}_Y} \mathcal{O}_X$.

Now letting X be a subscheme of $\mathbb{P}^n$, we define $\mathcal{O}_X(1)$ to be $i^*(\mathcal{O}_{\mathbb{P}^n}(1))$, where $i : X \to \mathbb{P}^n$ is the inclusion.

2.7.2 *Locally free sheaves*

A sheaf F of $\mathcal{O}_X$-modules is said to be *free* if F is isomorphic to a direct sum of copies of $\mathcal{O}_X$, and the number of copies of $\mathcal{O}_X$ is called its *rank*. The sheaf F is said to be *locally free* if X can be covered by open sets U such that $F|_U$ is a free $\mathcal{O}_U$-module (here $\mathcal{O}_U$ is just $\mathcal{O}_X|_U$). If the rank of (the locally free sheaf) F on any such U is the same, say n, then F is said to be a *locally free sheaf of rank n on X*.

2.7.3 *The scheme $V(\Omega)$ associated to a rank n locally free sheaf Ω*

Let $(X, \mathcal{O}_X)$ be a scheme. Let Ω be a locally free sheaf of rank n on X. Let $\{X_\alpha \mid \alpha \in I\}$ be an affine open cover of X, where $X_\alpha = \mathrm{Spec}(A_\alpha)$ for each $\alpha \in I$. Let B_α be the symmetric algebra of the A_α-module $\Omega(X_\alpha)$. Let $Y_\alpha = \mathrm{Spec}(B_\alpha)$, and $f_\alpha : Y_\alpha \to X_\alpha$ the canonical morphism. For every pair $\alpha, \beta \in I$, we have canonical isomorphisms $f_\alpha^{-1}(X_\alpha \cap X_\beta) \xrightarrow{\sim} f_\beta^{-1}(X_\alpha \cap X_\beta)$ (by the compatibility conditions on the affine covering $\{X_\alpha, \alpha \in I\}$). Hence the family $\{Y_\alpha, \alpha \in I\}$ can be glued together to define a scheme $(Y, \mathcal{O}_Y)$, and a morphism $f : Y \to X$ such that $f|_{Y_\alpha} = f_\alpha$.

Remark 2.7.4. The scheme Y is unique (up to isomorphism), i.e., if (Y', f') is another such pair, then there exists an isomorphism $\varphi : Y \to Y'$ such that the following diagram is commutative.

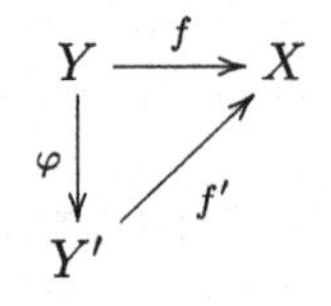

(*)

The scheme Y is denoted by $V(\Omega)$.

For $U = \mathrm{Spec}(A)$ open affine in X, we have $f^{-1}(U) \cong \mathrm{Spec}A[x_1, \ldots, x_n]$. We shall denote this isomorphism by ψ_U.

2.7.4 *Vector bundles*

Let $(X, \mathcal{O}_X)$ be a scheme. A (geometric) *vector bundle* of rank n over X is a scheme Y, and a morphism $f : Y \to X$, together with additional data consisting of an open covering $\{U_i\}$ of X, and isomorphisms $\psi_i : f^{-1}(U_i) \xrightarrow{\sim} \mathrm{Spec}A_i[x_1, \ldots, x_n]$ (here $\mathrm{Spec}A_i = U_i$), such that for any i, j and for any open affine subset $V = \mathrm{Spec}(A) \subset U_i \cap U_j$, the automorphism $\psi_j \circ \psi_i^{-1}$ of $\mathrm{Spec}A[x_1, \ldots, x_n]$ is given by a linear automorphism θ of $A[x_1, \ldots, x_n]$, i.e., $\theta(a) = a$ for all $a \in A$, and $\theta(x_i) = \sum a_{ij}x_j$ for suitable $a_{ij} \in A$.

An *isomorphism of two rank n vector bundles over X*

$$g : \{Y, f, \{U_i\}, \{\psi_i\}\} \longrightarrow \{Y', f', \{U_i'\}, \{\psi_i'\}\}$$

is an isomorphism $g : Y \to Y'$ such that $f = f' \circ g$ and such that Y, f, together with the covering of X consisting of the entire set $\{U_i\} \cup \{U_i'\}$, and the isomorphisms ψ_i and $\psi_i' \circ g$, is also a vector bundle on X.

Proposition 2.7.5 (cf. Chap. II, Ex. 5.18 of [28]). *Let Ω be a locally free sheaf of rank n on X. Let $\{U_i,\ i \in I\}$ be an affine covering of X such that $\Omega|_{U_i}$ is free (of rank n). Let $U_i = Spec(A_i)$, and $\psi_i : f^{-1}(U_i) \to SpecA_i[x_1, \ldots, x_n]$ the corresponding isomorphism. Then $\{Y, f, \{U_i\}, \{\psi_i\}\}$ is a vector bundle of rank n over X.*

Conversely, let $f : Y \to X$ be a rank n vector bundle over X. Let $\mathcal{S}_f$ be the presheaf given by $\mathcal{S}_f(U) = \{s : U \to Y$ morphism such that $f \circ s = Id_U\}$. Then $\mathcal{S}_f$ is in fact a locally free sheaf of $\mathcal{O}_X$-modules of rank n.

As a consequence of Proposition 2.7.5, we have the following.

Theorem 2.7.6. *There is a one-to-one correspondence between isomorphism classes of locally free sheaves of rank n on X, and isomorphism classes of rank n vector bundles over X.*

In view of the above bijection, we use the words "locally free sheaf" and "vector bundles" interchangeably.

Given a vector bundle Ω over X, we shall denote the *space of sections of Ω over* X by $H^0(X, \Omega)$, namely

$$H^0(X, \Omega) = \{s : X \to V(\Omega) \mid f \circ s = Id_X\}$$

(here, f is as in the diagram (*) above).

Definition 2.7.7. An *invertible sheaf* on X is a locally free sheaf of rank 1. The associated rank 1 vector bundle is called a *line bundle*.

2.8 Attributes of Varieties

2.8.1 Dimension of a topological space

For a topological space X, $\dim X$ is defined as the supremum of the lengths n of chains $F_0 \subset F_1 \subset \ldots \subset F_n$ of distinct irreducible closed sets in X.

If $X = X_1 \cup \ldots \cup X_r$, the X_i's being the irreducible components of X, then we have $\dim X = \max\{\dim X_i\}$.

Let $X = \mathrm{Spec}(A)$. Then clearly $\dim X = \dim A$ (where $\dim A$ is the Krull dimension of A [cf. Definition 2.1.7]). If in addition A is an integral domain and a finitely generated K-algebra, then $\dim X = \mathrm{tr.deg}_K K(A)$ $(= \mathrm{tr.deg}_K A)$, where $K(A)$

is the field of fractions of A. More generally, if X is an irreducible variety with function field $K(X)$, then $\dim X = \mathrm{tr.deg}_K K(X)$.

Example 2.8.1. We have $\dim \mathbb{A}^n = n$, $\dim \mathbb{P}^n = n$.

Example 2.8.2. If X is irreducible, then for any affine open subset U, $\dim U = \dim X$ (since $K(U) = K(X)$).

Proposition 2.8.3 (cf. [28]). *Let X be an irreducible variety, and Y a proper, closed subset. Then* $\dim Y < \dim X$.

The *codimension* $\mathrm{codim}_X Y$ of Y in X is defined to be $\dim X - \dim Y$.

2.8.2 Geometric properties of varieties

A point x on a variety X is said to be *normal* on X, if the local ring $\mathcal{O}_{X,x}$ is normal, i.e., $\mathcal{O}_{X,x}$ is an integral domain integrally closed in its quotient field. A variety X is said to be *normal* if every point x of X is normal on X. A projective variety X is *arithmetically normal* if the cone $\hat{X}$ is normal.

A variety X is said to be *Cohen–Macaulay* if $\mathcal{O}_{X,x}$ is Cohen–Macaulay for all $x \in X$ (i.e., $\mathrm{depth}\mathcal{O}_{X,x} = \dim \mathcal{O}_{X,x}$). A projective variety X is *arithmetically Cohen–Macaulay* if the cone $\hat{X}$ is Cohen–Macaulay.

A variety X is *factorial* if $\mathcal{O}_{X,x}$ is a factorial ring for all $x \in X$. A projective variety $X \hookrightarrow P^n$ is *arithmetically factorial* if the cone $\hat{X}$ is factorial.

A point x on a variety is called a *simple* or *smooth* point if $\mathcal{O}_{X,x}$ is a regular local ring, and X is said to be *nonsingular* or *smooth* if every point x of X is a smooth point. A point that is not smooth is called *singular*.

Definition 2.8.4. The *singular locus* of X is defined as

$$\mathrm{Sing}\, X := \{x \in X \mid x \text{ is singular}\}.$$

The singular locus is a closed subset of X; X is *nonsingular* when Sing X is empty.

2.8.3 The Zariski tangent space

Let x be a point on the variety X. Let $K(x) = \mathcal{O}_{X,x}/\mathfrak{m}_x$ be the residue field of the local ring $\mathcal{O}_{X,x}$ (viewed as an $\mathcal{O}_{X,x}$-module). The *Zariski tangent space* of X at x is the space $T_xX := \mathrm{Der}_K\,(\mathcal{O}_{X,x}, K(x))$, i.e., the space

$$\{D : \mathcal{O}_{X,x} \to K(x),\ K\text{-linear such that } D(ab) = D(a)b + aD(b)\}.$$

This is canonically isomorphic to $(\mathfrak{m}_x/\mathfrak{m}_x^2)^*$, the linear dual of $\mathfrak{m}_x/\mathfrak{m}_x^2$.

Remark 2.8.5. We have that $\dim T_xX \geq \dim X$ with equality if and only if x is a smooth point.

2.8.4 *The differential* $(d\phi)_x$

Given a morphism $\phi : X \to Y$ of varieties, the comorphism $\phi^* : \mathcal{O}_{Y,\phi(x)} \to \mathcal{O}_{X,x}$ induces a natural map $(d\phi)_x : T_xX \to T_{\phi(x)}Y$, called the *differential* of ϕ at x.

Chapter 3
Cohomology Theory

In this chapter, we do a brief recollection of the basics of cohomology theory leading up to the definition of sheaf cohomology.

3.1 Introduction to Category Theory

In this section, we recall the basics of category theory. For a more thorough introduction, we recommend [58, I, §11], [21, A5].

Definition 3.1.1. A *category* $\mathcal{A}$ is a collection of objects (denoted $\mathrm{Obj}\mathcal{A}$), and for each pair $E, F \in \mathrm{Obj}\mathcal{A}$ a set of morphisms (denoted $\mathrm{Mor}(E, F)$) with a composition law for three objects $E, F, G \in \mathrm{Obj}\mathcal{A}$:

$$\mathrm{Mor}(E, F) \times \mathrm{Mor}(F, G) \to \mathrm{Mor}(E, G), \quad (f, g) \mapsto g \circ f.$$

The composition law is associative, i.e., $(f \circ g) \circ h = f \circ (g \circ h)$. Additionally, for each $E \in \mathrm{Obj}\mathcal{A}$, there exists $Id_E \in \mathrm{Mor}(E, E)$ such that $Id_E \circ f = f$ (respectively, $g \circ Id_E = g$) for every $f \in \mathrm{Mor}(F, E)$ (resp., $g \in \mathrm{Mor}(E, F)$).

Definition 3.1.2. $\Phi : \mathcal{A}_1 \to \mathcal{A}_2$ is a *covariant functor* if

1. Φ maps objects of $\mathcal{A}_1$ to objects of $\mathcal{A}_2$,
2. Φ maps $\mathrm{Mor}(A, B) \to \mathrm{Mor}(\Phi(A), \Phi(B))$ such that
   a. $\Phi(Id_A) = Id_{\Phi(A)}$, for all $A \in \mathrm{Obj}(\mathcal{A}_1)$;
   b. $\Phi(g \circ f) = \Phi(g) \circ \Phi(f)$ whenever $g \circ f$ is defined.

V. Lakshmibai, J. Brown, *The Grassmannian Variety*,
Developments in Mathematics 42, DOI 10.1007/978-1-4939-3082-1_3

Similarly, $\Phi : \mathcal{A}_1 \to \mathcal{A}_2$ is a *contravariant functor* if the direction of morphisms is reversed, i.e.

1. Φ maps objects of $\mathcal{A}_1$ to objects of $\mathcal{A}_2$,
2. Φ maps $\mathrm{Mor}(A, B) \to \mathrm{Mor}(\Phi(B), \Phi(A))$ such that

   a. $\Phi(Id_A) = Id_{\Phi(A)}$, for all $A \in \mathrm{Obj}(\mathcal{A}_1)$;
   b. $\Phi(g \circ f) = \Phi(f) \circ \Phi(g)$ whenever $g \circ f$ is defined.

Given objects $X, Y \in \mathrm{Obj}\mathcal{A}$ for a category $\mathcal{A}$, we will define a categorical product and coproduct of X and Y.

Definition 3.1.3. A *categorical product* of X and Y is a triple (Z, p, v) where $Z \in \mathrm{Obj}\mathcal{A}$, $p \in \mathrm{Mor}(Z, X)$, and $v \in \mathrm{Mor}(Z, Y)$ such that given another triple (W, r, s) where $W \in \mathrm{Obj}\mathcal{A}$, $r \in \mathrm{Mor}(W, X)$, and $s \in \mathrm{Mor}(W, Y)$, then there exists a unique morphism $f : W \to Z$ such that $r = p \circ f$ and $s = v \circ f$.

We define the dual notion of *coproduct* of X and Y as follows: the coproduct of X and Y is a triple (Z, p, v) where $Z \in \mathrm{Obj}\mathcal{A}$, $p \in \mathrm{Mor}(X, Z)$, and $v \in \mathrm{Mor}(Y, Z)$ such that given another triple (W, r, s) where $W \in \mathrm{Obj}\mathcal{A}$, $r \in \mathrm{Mor}(X, W)$, and $s \in \mathrm{Mor}(Y, W)$, then there exists a unique morphism $f : Z \to W$ such that $r = f \circ p$ and $s = f \circ v$.

A *chain complex* in a category $\mathcal{A}$ is a sequence of objects and morphisms

$$\cdots \to A_{i+1} \stackrel{d_{i+1}}{\to} A_i \stackrel{d_i}{\to} A_{i-1} \to \cdots \to A_1 \stackrel{d_1}{\to} A_0 \to 0,$$

such that $d_i \circ d_{i+1} = 0$ for all $i \geq 1$. A *cochain complex*, on the other hand, is a sequence of objects and morphisms

$$0 \to A_0 \stackrel{d_0}{\to} A_1 \stackrel{d_1}{\to} A_2 \to \cdots \to A_{i-1} \stackrel{d_{i-1}}{\to} A_i \stackrel{d_i}{\to} A_{i+1} \to \cdots$$

such that $d_i \circ d_{i-1} = 0$ for all $i \geq 0$.

3.2 Abelian Categories

Throughout, let $\mathcal{A}$ be a category such that $\mathrm{Mor}(E, F)$ is an Abelian group for each pair of objects E, F. Let $E, F \in \mathrm{Obj}\mathcal{A}$, and $f \in \mathrm{Mor}(E, F)$. A *kernel* of f, denoted $\ker f$, is a pair (i, E'), where $i : E' \to E$ such that for all $M \in \mathrm{Obj}\mathcal{A}$, the sequence

$$0 \to \mathrm{Mor}(M, E') \to \mathrm{Mor}(M, E) \to \mathrm{Mor}(M, F)$$

is exact. In other words, given $\theta : M \to E$, the composite $f \circ \theta$ is the zero map if and only if there exists $\theta' : M \to E'$ such that $\theta = i \circ \theta'$. One may verify that a kernel of f is unique up to isomorphism.

On the other hand, a *cokernel* of f, denoted coker$f$, is a pair (i', F'), where $i' : F \to F'$ is a morphism such that for all $M \in \text{Obj}\mathcal{A}$, the sequence

$$0 \to \text{Mor}(F', M) \to \text{Mor}(F, M) \to \text{Mor}(E, M)$$

is exact. In other words, given $\delta : F \to M$, the composite $\delta \circ f$ is the zero map if and only if there exists $\delta' : F' \to M$ such that $\delta = \delta' \circ i'$. As above, a cokernel of f is unique up to isomorphism.

Definition 3.2.1. Let $\mathcal{A}$ be a category, $\mathcal{A}$ is an *Abelian category* if

1. For each pair of objects E, F, $\text{Mor}(E, F)$ is an Abelian group.
2. The law of composition of morphisms

$$\text{Mor}(E, F) \times \text{Mor}(F, G) \to \text{Mor}(E, G)$$
$$(f, g) \mapsto g \circ f$$

   is bilinear.
3. There exists a zero object 0 such that $\text{Mor}(0, E)$ and $\text{Mor}(E, 0)$ have precisely one element, for all E.
4. Products and coproducts exist in $\mathcal{A}$.
5. Kernels and cokernels of morphisms exist in $\mathcal{A}$.
6. For $f \in \text{Mor}(E, F)$,

   a. If ker$f$ is the zero object, then (f, E) is the kernel of $i' : F \to F'$; i.e., f is the kernel of its cokernel.
   b. If coker$f$ is the zero object, then (f, F) is the cokernel of $i : E' \to E$; i.e., f is the cokernel of its kernel.
   c. If both ker$f$ and coker$f$ are the zero object, then f is an isomorphism.

A category that has properties 1–4 above is called an *additive category*.

Example 3.2.2. The following are some examples of Abelian categories:

1. The category of Abelian groups.
2. The category of A-modules for a ring A.
3. The category of sheaves of Abelian groups on a topological space.
4. The category of sheaves of $\mathcal{O}_X$-modules on a ringed space.

Definition 3.2.3. Let $\Phi : \mathcal{A}_1 \to \mathcal{A}_2$ be a functor of Abelian categories, Φ is an *additive functor* if the map $\text{Mor}(A, B) \to \text{Mor}(\Phi(A), \Phi(B))$ is a homomorphism of Abelian groups for all $A, B \in \text{Obj}\mathcal{A}_1$.

Definition 3.2.4. Let $\Phi : \mathcal{A}_1 \to \mathcal{A}_2$ be an additive covariant functor of Abelian categories. We say Φ is *left exact* if given an exact sequence of objects in $\mathcal{A}_1$

$$0 \to A_1 \to A_2 \to A_3,$$

the sequence

$$0 \to \Phi(A_1) \to \Phi(A_2) \to \Phi(A_3)$$

is exact. We say Φ is *right exact* if given the exact sequence of objects in $\mathcal{A}_1$

$$A_1 \to A_2 \to A_3 \to 0,$$

the sequence

$$\Phi(A_1) \to \Phi(A_2) \to \Phi(A_3) \to 0$$

is exact.

Remark 3.2.5. If $\Phi : \mathcal{A}_1 \to \mathcal{A}_2$ is a contravariant functor, one has a similar definition of right and left exact. Namely, Φ is left exact if given the exact sequence

$$A_1 \to A_2 \to A_3 \to 0,$$

the sequence

$$0 \to \Phi(A_3) \to \Phi(A_2) \to \Phi(A_1)$$

is exact. We say Φ is right exact if given the exact sequence

$$0 \to A_1 \to A_2 \to A_3,$$

the sequence

$$\Phi(A_3) \to \Phi(A_2) \to \Phi(A_1) \to 0$$

is exact.

Example 3.2.6. Let A be a ring, and let $\mathcal{A}_1 = \mathcal{A}_2$ be the category of A-modules. We fix an A-module M, and define the functor $\Phi = \mathrm{Hom}(M, \cdot)$, where $\Phi(N) = \mathrm{Hom}(M,N)$. Note that $\mathrm{Hom}(M,N)$ is an A-module, where for $a \in A$ and $\phi \in \mathrm{Hom}(M,N)$, $a\phi(m) = \phi(am)$. We also note that Φ maps $\mathrm{Mor}(N_1,N_2)$ to $\mathrm{Mor}(\mathrm{Hom}(M,N_1),\mathrm{Hom}(M,N_2))$ by composition. Thus $\Phi : \mathcal{A}_1 \to \mathcal{A}_2$ is a covariant functor, and one can verify that Φ is left exact.

Remark 3.2.7. More generally, for any Abelian category $\mathcal{A}_1$ and $M \in \mathrm{Obj}\mathcal{A}_1$, the map given by $\mathrm{Mor}(M, \cdot)$ is a left exact functor from $\mathcal{A}_1$ to the category of Abelian groups.

Example 3.2.8. Let A be a ring, and let $\mathcal{A}_1 = \mathcal{A}_2$ be the category of A-modules. We fix an A-module M, and define the functor Φ such that $\Phi(N) = N \otimes_A M$. Then Φ is a right exact functor.

Definition 3.2.9. Let $\mathcal{A}$ be an Abelian category, $M \in \mathrm{Obj}\mathcal{A}$ is *projective* if the covariant functor $\mathrm{Hom}(M, \cdot)$ is an exact functor. Equivalently, given $N \overset{\delta}{\to} P \to 0$ exact, $\mathrm{Hom}(M,N) \to \mathrm{Hom}(M,P) \to 0$ is exact; i.e., for a morphism $\theta : M \to P$, there exists $\phi : M \to N$ such that $\theta = \delta \circ \phi$.

Similarly, M is *injective* if the contravariant functor $\text{Hom}(\cdot, M)$ is an exact functor. Equivalently, given $0 \to P \xrightarrow{\delta} N$ exact, $\text{Hom}(N, M) \to \text{Hom}(P, M) \to 0$ is exact; i.e., for a morphism $\theta : P \to M$, there exists $\phi : N \to M$ such that $\theta = \phi \circ \delta$.

Definition 3.2.10. An Abelian category $\mathcal{A}$ has *enough injectives* if given $F \in \text{Obj}\mathcal{A}$, there exists an injective object $I \in \text{Obj}\mathcal{A}$ and a monomorphism

$$0 \to F \xrightarrow{i} I$$

We note that in an Abelian category with enough injectives, any object may be regarded as a subobject of an injective object.

Proposition 3.2.11. *If $\mathcal{A}$ has enough injectives, then each object has an injective resolution; in other words, for all $A \in \text{Obj}\mathcal{A}$, there exists an exact sequence*

$$0 \to A \to I_0 \to I_1 \to I_2 \to \dots,$$

such that I_j is injective for all $j \geq 0$.

Proof. By the definition of enough injectives, there exists an injective object I_0 such that $0 \to A \xrightarrow{i} I_0$ is exact. Let $A_1 = I_0/A$ (i.e., let $A_1 = \text{coker}\, i$). Because $\mathcal{A}$ has enough injectives, there exists an injective object I_1 and a monomorphism $i_1 : A_1 \to I_1$. By projecting I_0 onto A_1, we have the exact sequence

$$0 \longrightarrow A \longrightarrow I_0 \longrightarrow I_1.$$

Thus proceeding, we obtain an injective resolution as desired. □

3.2.1 Derived Functors

Let F be a left exact, additive, covariant functor of Abelian categories, $F : \mathcal{A}_1 \to \mathcal{A}_2$, where $\mathcal{A}_1$ has enough injectives. Let $A \in \text{Obj}\mathcal{A}_1$, and take an injective resolution of A:

$$0 \to A \longrightarrow I_0 \xrightarrow{d_0} I_1 \longrightarrow \dots$$

We apply F to the injective resolution (omitting A) to obtain a cochain complex

$$C^{\cdot} : \quad 0 \longrightarrow F(I_0) \xrightarrow{F(d_0)} F(I_1) \longrightarrow \dots$$

Note that because F is additive, the zero map in $\text{Mor}(I_{j-1}, I_{j+1})$ is mapped to the zero map in $\text{Mor}(F(I_{j-1}), F(I_{j+1}))$.

Definition 3.2.12. We define R^iF, the *right derived functor* associated to F as

$$R^iF(A) = H^i(C^\cdot) \left(= \frac{\mathrm{ker} F(d_i)}{\mathrm{Im} F(d_{i-1})} \right).$$

Remark 3.2.13.

1. The definition of $R^iF(A)$ is independent of the choice of the injective resolution of A.
2. $R^0F(A) = F(A)$. This is because $R^0F(A) = \ker F(d_0)$, and because F is a left exact functor, $\ker F(d_0) = F(A)$.
3. $R^iF : \mathcal{A}_1 \to \mathcal{A}_2$ is an additive functor.
4. Given a short exact sequence

$$0 \longrightarrow A' \longrightarrow A \longrightarrow A'' \longrightarrow 0,$$

there exists a natural set of morphisms $\delta^i : R^iF(A'') \to R^{i+1}F(A')$, $i \geq 0$ such that the sequence

$$\cdots \to R^iF(A') \to R^iF(A) \to R^iF(A'') \xrightarrow{\delta^i} R^{i+1}F(A') \to \cdots$$

is exact. (This is a result of the "snake lemma," cf. [58, III, §9] for example.)
5. Given a morphism of two short exact sequences

$$\begin{array}{ccccccccc} 0 & \longrightarrow & A' & \longrightarrow & A & \longrightarrow & A'' & \longrightarrow & 0 \\ & & \downarrow{\scriptstyle f'} & & \downarrow{\scriptstyle f} & & \downarrow{\scriptstyle f''} & & \\ 0 & \longrightarrow & B' & \longrightarrow & B & \longrightarrow & B'' & \longrightarrow & 0 \end{array}$$

The following diagram is commutative:

$$\begin{array}{ccc} R^iF(A'') & \xrightarrow{\delta^i} & R^{i+1}F(A') \\ \downarrow{\scriptstyle f''} & & \downarrow{\scriptstyle f'} \\ R^iF(B'') & \xrightarrow{\delta^i} & R^{i+1}F(B') \end{array}$$

6. For an injective object I, $R^iF(I) = 0$ for $i \geq 1$, because one may work with the injective resolution $0 \to I \to I \to 0$.

3.3 Enough Injective Lemmas

In this section, we prove that certain categories have enough injectives.

Lemma 3.3.1 (Baer's Lemma). *Let Q be an R-module such that for every*

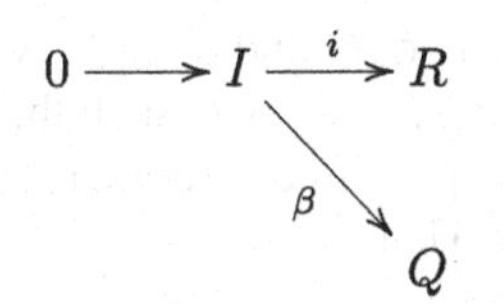

where I is an ideal of R, there exists $\gamma : R \to Q$ such that $\beta = \gamma \circ i$. Then Q is injective in the category of R-modules.

Proof. We begin with the following diagram of R-modules:

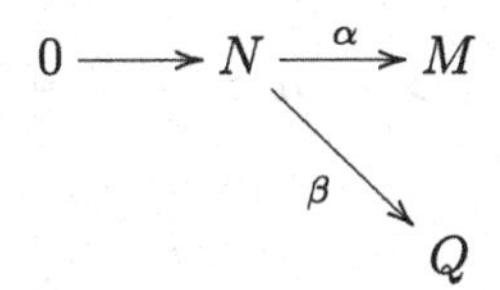

To show that Q is injective, we must show that β extends to M.

Let $\mathcal{M} = \{(\beta', N') \mid N \subseteq N' \subseteq M,\ \beta' : N' \to Q \text{ such that } \beta = \beta' \circ \alpha\}$. Note that $\mathcal{M}$ is nonempty, since $(\beta, N) \in \mathcal{M}$. We define a partial order on $\mathcal{M}$:

$$(\beta_1, N_1) < (\beta_2, N_2), \text{ if } N_1 \subset N_2 \text{ and } \beta_2|_{N_1} = \beta_1.$$

Now if we have a totally ordered subset of $\mathcal{M}$: $\{(\beta_j, N_j)\}$, we can see that this set has a maximal element by defining $\bar{N} = \sum N_j$ and $\bar{\beta}(x) = \beta_j(x)$ if $x \in N_j$; then $(\bar{\beta}, \bar{N}) \in \mathcal{M}$. Thus, by Zorn's Lemma, there exists a maximal submodule N' of M and an extension β'.

We now claim that $N' = M$, the proof of which will complete the proof of the lemma. Assume $N' \subsetneq M$. Choose $m \in M \setminus N'$, let $N'' = N' + Rm$. Let $I = \{r \in R \mid rm \in N'\}$. By hypothesis, the map $I \to Q$ such that $r \mapsto \beta'(rm)$ extends to a map $\delta : R \to Q$. Define $\delta' : Rm \to Q$ as $\delta'(rm) = \delta(r)$; note that if $r_1 m = r_2 m$, then $r_1 - r_2$ is in I, and hence $\delta(r_1 - r_2) = \beta'((r_1 - r_2)m) = 0$, and hence $\delta(r_1) = \delta(r_2)$. Thus δ' is well defined. Further, if $rm \in N'$, then $r \in I$, and hence $\delta'(rm) = \delta(r) = \beta'(rm)$. Thus we have $\delta' = \beta'$ on $Rm \cap N'$.

Therefore, we may define an extension β'' of β' to N'', by letting

$$\beta''(x) = \begin{cases} \beta'(x) & \text{if } x \in N' \\ \delta'(x) & \text{if } x \in Rm. \end{cases}$$

Hence $(\beta'', N'') \in \mathcal{M}$. This contradicts the maximality of (β', N'). Thus $N' = M$, and the result follows. □

Definition 3.3.2. Let G be an Abelian group, G is *divisible* if for every $g \in G$, and for every nonzero integer n, there exists $g' \in G$ such that $ng' = g$.

Example 3.3.3. Both $\mathbb{Q}/\mathbb{Z}$ and $\mathbb{R}/\mathbb{Z}$ are divisible groups.

Lemma 3.3.4. *Let G be an object of the category of Abelian groups. G is injective if and only if G is divisible.*

Proof. Let G be injective. Let $g \in G$ and $n \in \mathbb{Z}$, we will show that there exists $g' \in G$ such that $g = ng'$. Define $\beta : \mathbb{Z} \to G$ such that $\beta(1) = g$, and $\alpha_n : \mathbb{Z} \to \mathbb{Z}$ such that $\alpha_n(t) = tn$. Because G is injective, there exists $\gamma : \mathbb{Z} \to G$ such that $\beta = \gamma \circ \alpha_n$. Denote $g' = \gamma(1)$. Then

$$g = \beta(1) = \gamma(\alpha_n(1)) = n\gamma(1) = ng'.$$

Therefore G is divisible.

Now let G be divisible. Let $n \in \mathbb{Z}$, and suppose we have the following

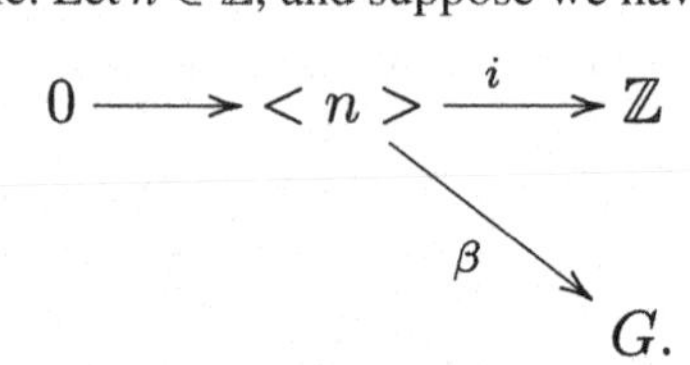

We must show that there exists $\gamma : \mathbb{Z} \to G$ that completes the diagram and makes it a commutative diagram; then by Baer's Lemma we will have that G is injective (note that Abelian groups are $\mathbb{Z}$-modules). Let $g = \beta(n)$. Because G is divisible, there exists g' such that $ng' = g$. Define $\gamma : \mathbb{Z} \to G$ such that $\gamma(1) = g'$. Then $\gamma(n) = ng' = g = \beta(n)$, and therefore we have $\beta = \gamma \circ i$. The result follows. □

Proposition 3.3.5. *The category of Abelian groups has enough injectives.*

Proof. Let M be a $\mathbb{Z}$-module. We need to find an injective object and an injective map from M to that object. Denote $M^\vee = \mathrm{Hom}_\mathbb{Z}(M, \mathbb{Q}/\mathbb{Z})$, and define a map $i : M \to (M^\vee)^\vee$, $m \mapsto \phi_m : M^\vee \to \mathbb{Q}/\mathbb{Z}$, $\phi_m(f) = f(m)$. The map i is in fact an injective map (which can be seen by tensoring with $\mathbb{Q}$ and using the fact that for a vector space V, $(V^\vee)^\vee = V$).

Now represent $M^\vee$ as a quotient of a free Abelian group F:

$$F \to M^\vee \to 0.$$

Since the contravariant functor $\mathrm{Hom}(\cdot, \mathbb{Q}/\mathbb{Z})$ is left exact, this induces

$$0 \to (M^\vee)^\vee \xrightarrow{j} F^\vee.$$

On the other hand, F being a free Abelian group, $F^\vee = \mathrm{Hom}_\mathbb{Z}(F, \mathbb{Q}/\mathbb{Z})$ is a direct product of groups isomorphic to $\mathbb{Q}/\mathbb{Z}$ and therefore $F^\vee$ is divisible (because $\mathbb{Q}/\mathbb{Z}$ is divisible). Hence, by Lemma 3.3.4, $F^\vee$ is injective in the category of Abelian groups and the result follows from the composition of injective maps

$$M \xrightarrow{i} (M^\vee)^\vee \xrightarrow{j} F^\vee.$$

□

Lemma 3.3.6. *Let A be an S-algebra, and let Q' be an injective S-module. Then $\mathrm{Hom}_S(A, Q')$ is an injective A-module.*

Proof. Let us denote $\mathrm{Hom}_S(A, Q')$ by Q. To begin, we provide an A-module structure for Q. Given an S-linear map $f : A \to Q'$ and $a, x \in A$, define

$$(a \cdot f)(x) = f(ax).$$

Now we show that Q is an injective object in the category of A-modules. Given

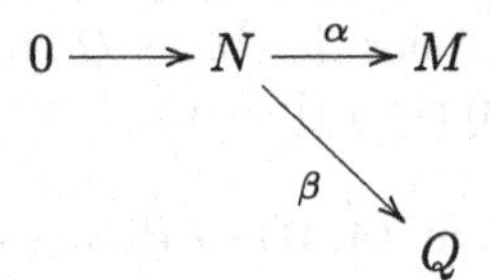

where β is A-linear, we need to find $\gamma : M \to Q$ that is A-linear, completes the diagram and makes it a commutative diagram. First, define $\theta : Q \to Q'$ such that for $\phi \in Q$, $\theta(\phi) = \phi(1) \in Q'$. Now define $\beta' : N \to Q'$ as $\beta' = \theta \circ \beta$. Because Q' is an injective S-module, there exists an S-linear map γ' that completes the diagram

$$\begin{array}{ccccc} 0 & \longrightarrow & N & \xrightarrow{\alpha} & M \\ & & & \underset{\beta'}{\searrow} & \downarrow \gamma' \\ & & & & Q' \end{array}$$

and makes it a commutative diagram.

Define $\gamma : M \to Q$:

$$\gamma(m) : A \to Q'$$
$$a \mapsto \gamma'(am)$$

To see that γ is A-linear, recall the A-module structure defined above; thus

$$a \cdot (\gamma(m))(x) = (\gamma(m))(ax) = \gamma'(axm).$$

On the other hand,

$$(\gamma(am))(x) = \gamma'(xam).$$

Therefore $\gamma(am) = a\gamma(m)$.

We have that $\theta \circ \gamma \circ \alpha = \theta \circ \beta$, and we want to show that $\gamma \circ \alpha = \beta$. Let $n \in N$ and $a \in A$. We have

$$(\gamma \circ \alpha\,(n))\,(a) = (\gamma\,(\alpha\,(n)))\,(a) = \gamma'\,(a\alpha\,(n)) = \gamma' \circ \alpha(an)$$

since α is A-linear. This is then equal to

$$\beta'(an) = \theta \circ \beta(an) = (\beta\,(an))\,(1) = (a\beta\,(n))\,(1)$$

since β is A-linear. Finally, by the definition of the A-module structure on Q, this equals $(\beta\,(n))\,(a)$. Hence $\gamma \circ \alpha = \beta$. □

Proposition 3.3.7. *Let A be a ring, then the category of A-modules has enough injectives.*

Proof. Let M be an A-module, and we begin with the map $\alpha : M \to \mathrm{Hom}_{\mathbb{Z}}(A, M)$ such that $m \mapsto \alpha_m : A \to M$, $\alpha_m(a) = am$. We note that $\alpha_m(1) = m$, for all $m \in M$, and thus α is an injective map.

Viewing M as an Abelian group, we have from Proposition 3.3.5, there exists a monomorphism of Abelian groups $\beta : M \to Q'$, for Q' some injective Abelian group. Applying the left exact functor $\mathrm{Hom}_{\mathbb{Z}}(A, \cdot)$, we have

$$0 \to \mathrm{Hom}_{\mathbb{Z}}(A, M) \to \mathrm{Hom}_{\mathbb{Z}}(A, Q')$$

is exact. We can now apply Lemma 3.3.6 (with $S = \mathbb{Z}$) to say that $\mathrm{Hom}_{\mathbb{Z}}(A, Q')$ is an injective A-module, and since α as described above is injective,

$$0 \to M \to \mathrm{Hom}_{\mathbb{Z}}(A, Q')$$

is exact. Therefore the category of A-modules has enough injectives. □

For a topological space X, let $\mathcal{A}b(X)$ denote the category of sheaves of Abelian groups on X. For a ringed space $(X, \mathcal{O}_X)$, let $\mathcal{M}odX$ denote the category of sheaves of $\mathcal{O}_X$-modules.

Theorem 3.3.8. *Let $(X, \mathcal{O}_X)$ be a ringed space, then $\mathcal{M}odX$ has enough injectives.*

Proof. Let $\mathcal{G}$ be a sheaf of $\mathcal{O}_X$-modules. Thus for each $x \in X$, $\mathcal{G}_x$ is an $\mathcal{O}_{X,x}$-module. By Proposition 3.3.7, there exists an injective $\mathcal{O}_{X,x}$-module I_x such that

$$0 \longrightarrow \mathcal{G}_x \longrightarrow I_x$$

is exact. Let $\mathcal{I}$ be the $\mathcal{O}_X$-module such that

$$\mathcal{I}(U) = \prod_{x \in U} I_x$$

for each open set $U \subseteq X$. We have

$$\mathcal{H}om_{\mathcal{O}_X}(\mathcal{G}, \mathcal{I}) = \prod_{x \in X} \mathrm{Hom}_{\mathcal{O}_{X,x}}(\mathcal{G}_x, I_x).$$

Hence the stalk maps $\mathcal{G}_x \to I_x$ (which are injective for every $x \in X$) induce an injective map $\mathcal{G} \to \mathcal{I}$.

If we show that $\mathcal{I}$ is an injective $\mathcal{O}_X$-module, we will be done. To do so, we will show that $\mathrm{Hom}_{\mathcal{O}_X}(\cdot, \mathcal{I})$ is exact. We first note that a sequence

$$0 \to \mathcal{F} \to \mathcal{L} \to \mathcal{H} \to 0$$

is exact if and only if the induced sequence

$$0 \to \mathcal{F}_x \to \mathcal{L}_x \to \mathcal{H}_x \to 0$$

is exact for all $x \in X$. Let us now begin with an exact sequence of $\mathcal{O}_X$-modules:

$$\begin{aligned}
&0 \to \mathcal{F} \to \mathcal{L} \to \mathcal{H} \to 0.\\
\Rightarrow\ &0 \to \mathcal{F}_x \to \mathcal{L}_x \to \mathcal{H}_x \to 0 \text{ is exact}, \forall\, x \in X.\\
\Rightarrow\ &0 \to \operatorname{Hom}(\mathcal{H}_x, I_x) \to \operatorname{Hom}(\mathcal{L}_x, I_x) \to \operatorname{Hom}(\mathcal{F}_x, I_x) \to 0
\end{aligned}$$

is exact because I_x is injective for all $x \in X$. Therefore

$$0 \to \operatorname{Hom}(\mathcal{H}, \mathcal{I}) \to \operatorname{Hom}(\mathcal{L}, \mathcal{I}) \to \operatorname{Hom}(\mathcal{F}, \mathcal{I}) \to 0$$

is exact, and the result follows. □

Corollary 3.3.9. *For X a topological space, the category $\mathcal{A}b(X)$ has enough injectives.*

Proof. Define $\mathcal{O}_X$ to be the constant sheaf of rings determined by $\mathbb{Z}$ (see Definition 2.7.3 letting $R = \mathbb{Z}$). Then $(X, \mathcal{O}_X)$ is a ringed space and $\mathcal{M}odX = \mathcal{A}b(X)$. □

3.4 Sheaf and Local Cohomology

For any topological space X, we define a functor Γ from $\mathcal{A}b(X)$ to the category of all Abelian groups: for $\mathcal{F} \in \mathcal{A}b(X)$, let $\Gamma(X, \mathcal{F}) = \mathcal{F}(X)$. This is called the *global sections functor*.

The functor Γ is left exact: given $0 \to \mathcal{F} \to \mathcal{G}$ that is exact, then $0 \to \Gamma(X, \mathcal{F}) \to \Gamma(X, \mathcal{G})$ is exact (because it is exact at the level of the stalks, and we are looking at sections $X \to \dot{\cup}\mathcal{F}_x \hookrightarrow \dot{\cup}\mathcal{G}_x$ of the map $\dot{\cup}\mathcal{F}_x \to X, a\, (\in \mathcal{F}_x) \mapsto x$).

On the other hand, Γ is not right exact; we provide an example to illustrate this. Let X be a connected T_1-space (namely, a point set $\{x\}$ is closed in X). Fix $x_1, x_2 \in X$. Define $\mathcal{F} \in \mathcal{A}b(X)$ such that

$$\mathcal{F}_x = \begin{cases} (0) & \text{if } x \notin \{x_1, x_2\} \\ K & \text{if } x \in \{x_1, x_2\}, \end{cases}$$

K being a field. (Such a sheaf is sometimes called the skyscraper sheaf.) Then $\Gamma(X, \mathcal{F}) = K \oplus K$. On the other hand, let $\mathcal{G}$ be the constant sheaf, $\mathcal{G}(U) = \{f : U \to K \text{ continuous}\}$ where K is given the discrete topology.

Thus, $\mathcal{G} \to \mathcal{F} \to 0$ is exact because at the level of stalks, $\mathcal{G}_x \to \mathcal{F}_x$ is surjective for all $x \in X$. On the other hand, $\Gamma(X, \mathcal{G}) \to \Gamma(X, \mathcal{F}) \to 0$ is not exact (since $\Gamma(X, \mathcal{G}) = K, \Gamma(X, \mathcal{F}) = K \oplus K$).

Thus, Γ is a left exact, additive, covariant functor, and $\mathcal{A}b(X)$ is an Abelian category with enough injectives; thus as in Definition 3.2.12, we define the right derived functor $R^i\Gamma : \mathcal{A}b(X) \to Ab$ (where Ab is the category of Abelian groups).

Definition 3.4.1. Given $\mathcal{F}$, a sheaf of Abelian groups on X, let $H^i(X, \mathcal{F}) = R^i\Gamma(X, \mathcal{F})$; $\{H^i(X, \mathcal{F}) \mid i \geq 0\}$ are defined as the *cohomology groups of* $\mathcal{F}$.

Next we give the definition of local cohomology. Let R be a ring, let I be an ideal of R, and let M be an R-module. We define the 0th local cohomology module of M with supports in I to be the set of elements in M that are annihilated by a power of I; i.e.

$$H_I^0(M) = \bigcup_{n \geq 1} \{m \in M \mid I^n m = 0\}.$$

To see that H_I^0 is left exact as a functor, suppose we have an exact sequence of R-modules:

$$0 \to M \xrightarrow{\phi} N,$$

where ϕ is R-linear. If $m \in M$ is annihilated by $r \in R$, and $\phi(m) = n$, then $\phi(rm) = r\phi(m) = rn$, and thus n is also annihilated by r. Therefore, the induced map $H_I^0(M) \to H_I^0(N)$ is injective.

Therefore we have H_I^0 is a left exact, additive, covariant functor. And as shown in Proposition 3.3.7, the category of R-modules has enough injectives. The higher local cohomology groups are defined using the right derived functor.

Definition 3.4.2. Given R, I, and M as above, let $H_I^i(M) = R^i H_I^0(M)$, $\{H_I^i(M) \mid i \geq 0\}$ are defined as the *local cohomology groups of* M with supports in I.

We cite the following theorem without proof:

Theorem 3.4.3 (cf. Theorem A4.3 [21]). *Let* $(R, \mathfrak{m})$ *be a local ring, and let* M *be a finitely generated* R*-module. Let* $d = \dim M$*, and* $\delta = \text{depth } M$*. Then*

1. $H_{\mathfrak{m}}^i(M) = 0$ *for* $i < \delta$, $i > d$, *and*
2. $H_{\mathfrak{m}}^i(M) \neq 0$ *for* $i = \delta$, $i = d$.

In view of Proposition 2.1.14, we have the following corollary.

Corollary 3.4.4. *For a local ring* $(R, \mathfrak{m})$ *and* R*-module* M*, we have*

$$\text{depth } M = \min\{i \mid H_{\mathfrak{m}}^i(M) \neq 0\}.$$

Chapter 4
Gröbner Bases

In this chapter, we present the basics on Gröbner bases, and apply it to flat degenerations.

4.1 Monomial Orders

Throughout this section, let S be the polynomial ring $K[x_1, \ldots, x_r]$, where K is a field. A *monomial* is an element of S of the form $x_1^{a_1} \ldots x_r^{a_r}$ for $a_i \in \mathbb{Z}_{\geq 0}$ for all $1 \leq i \leq r$; for $\underline{a} = (a_1, \ldots, a_r) \in \mathbb{Z}_{\geq 0}^r$; we will sometimes use the notation $x^{\underline{a}}$ in place of $x_1^{a_1} \ldots x_r^{a_r}$. We also note that the monomial $1 = x^{\underline{0}}$.

The monomial $x^{\underline{a}}$ is *divisible* by the monomial $x^{\underline{b}}$ if there exists an $f \in S$ such that

$$x^{\underline{a}} = fx^{\underline{b}}.$$

An ideal $I \subset S$ is a *monomial ideal* if I can be generated by a (finite) set of monomials in S. We note that given a set of monomial generators for I, deciding whether a given monomial $x^{\underline{a}} \in S$ is in I is as easy as seeing if $x^{\underline{a}}$ is divisible by any of the monomial generators of I. This property also simplifies working in S/I.

On the other hand, if J is any ideal of S, we would like to know what monomials remain linearly independent in S/J. If a monomial ideal I contains at least one monomial from each polynomial in J, then the set of all monomials of S not in I remain linearly independent in S/J. Of course, we would like to select the monomials that form the generators of I in such a way as to be minimal. This idea motivates the following definitions of monomial order and Gröbner basis.

Definition 4.1.1. A *monomial order* on S is a total order $>$ on the monomials such that for $x^{\underline{a}}, x^{\underline{b}} \in S$, and $1 \neq x^{\underline{m}} \in S$, we have:

$$x^{\underline{a}} > x^{\underline{b}} \text{ implies } x^{\underline{m}}x^{\underline{a}} > x^{\underline{m}}x^{\underline{b}} > x^{\underline{b}}.$$

V. Lakshmibai, J. Brown, *The Grassmannian Variety*,
Developments in Mathematics 42, DOI 10.1007/978-1-4939-3082-1_4

There are many different monomial orders. We give an example of one commonly used order:

Definition 4.1.2. Take a total order $x_1 > x_2 > \ldots > x_r$ on $\{x_1, \ldots, x_r\} \subset S$. The *lexicographic order* $>^{lex}$ on the monomials of S is given as follows: let $\underline{a} = (a_1, \ldots, a_r)$ and $\underline{b} = (b_1, \ldots, b_r)$. Then $x^{\underline{a}} >^{lex} x^{\underline{b}}$ if there exists $1 \leq s \leq r$ such that $a_1 = b_1, \ldots, a_{s-1} = b_{s-1}$ but $a_s > b_s$.

Example 4.1.3. For example, in $K[x_1, x_2, x_3, x_4]$,

$$x_1 x_4^4 >^{lex} x_2 x_3^2 x_4^2.$$

We leave it to the reader to check that $\geq^{lex}$ is in fact a monomial order.

Lemma 4.1.4. *A monomial order is Artinian, i.e., every nonempty subset of monomials has a least element.*

Proof. Let X be a set of monomials (not necessarily finite), and let I_X be the ideal generated by X. Since S is Noetherian, I_X is finitely generated, say $I_X =< f_1, \ldots, f_r >$ where $f_j = \sum_i a_{ij} m_{ij}$, where $a_{ij} \in K$, $m_{ij} \in X$. Thus $I_X =< m_{ij} >$. Let $Y = \{m_{ij}\}$, a finite collection. Let m be the least element of Y. Any monomial in I_X is of the form nm_{ij} for n a monomial in S and hence $nm_{ij} \geq m_{ij} > m$. Therefore m is the least element of X. □

Lemma 4.1.5. *Suppose we have an order $>$ on monomials of S such that $nm_1 > nm_2$ whenever $m_1 > m_2$ (without requiring that $nm_2 > m_2$). Then the condition that $nm > m$ for all monomials $n, m \in S$ is equivalent to the order being Artinian.*

Proof. First observe that under the hypothesis, the condition $nm > m$ is equivalent to the condition that $n > 1$ for all monomials $n \in S$. Let us show that $n > 1$ for all monomials $n \in S$ if and only if the order is Artinian. From the previous remark, we have that $n > 1$ implies the order is Artinian.

To prove the converse, let the order $>$ be Artinian. Thus there exists a smallest monomial in S, call it m. Assume $m \neq 1$, then $1 > m$. The hypothesis implies $m > m^2$, and m^2 is smaller than the smallest monomial m, a contradiction. □

Lemma 4.1.6. *For any order $>$ (not necessarily a monomial order) on the set M of monomials in S, the order is Artinian if and only if M satisfies the descending chain condition, i.e., any chain of elements of M, $m_1 > m_2 > \ldots$ is finite.*

Proof. Assume $>$ has the descending chain condition, but is not Artinian. This implies the existence of a nonempty set A of elements of M with no least element. Choose $m_1 \in A$, since it is not the least element, there exists $m_2 \in A$ such that $m_1 > m_2$. Continuing thus we arrive at an infinite sequence $m_1 > m_2 > \ldots$, a contradiction.

Conversely, assume $>$ is Artinian, but that there exists an infinite sequence $m_1 > m_2 > \ldots$. Then the set $\{m_1, m_2, \ldots\}$ has no least element, a contradiction. □

Definition 4.1.7. Given a monomial order $>$ and a polynomial $f \in S$, let $\text{in}_>(f)$ be the term of f greater than all other terms of f with respect to $>$, called the *initial term of f*. Let I be an ideal of S, we define $\text{in}_>(I)$ to be the monomial ideal generated by $\{\text{in}_>(f) \mid f \in I\}$, called the *initial ideal* of I.

Remark 4.1.8. By "term of f," we simply allow for the greatest monomial appearing in f to have any coefficient other than 0, i.e., the initial term need not have a coefficient equal to 1.

As a cautionary note, we point out that in general, $\text{in}_>(I)$ is not generated by the initial terms of a set of generators of I, as illustrated in the following example.

Example 4.1.9. For the order $y > x$ on $K[x, y]$, and $I =< f_1, f_2 >$, where $f_1 = xy, f_2 = x - y$, we have $\text{in}(f_1) = xy, \text{in}(f_2) = y$. Now $x^2 - xy (= x(x - y)) \in I$, and hence $x^2 \in I$, but $x^2 (= \text{in}(x^2))$ is not in the ideal generated by $\text{in}(f_1), \text{in}(f_2)$.

Remark 4.1.10. $\text{in}_>(fg) = (\text{in}_>(f))\,(\text{in}_>(g))$.

Theorem 4.1.11 (Macaulay). *Let I be an ideal in S and $>$ any monomial order. Then the set of all monomials of S not in $\text{in}_>(I)$ forms a basis of S/I as a K-vector space.*

Proof. Let M be the set of all monomials of S not in $\text{in}_>(I)$. Suppose we have a dependence relation on monomials of M in S/I:

$$f = \sum_{i=1}^{n} k_i \underline{s}_i$$

where $f \in I$, $0 \neq k_i \in K$, and $\underline{s}_i \in M$ for $1 \leq i \leq n$. Then $\text{in}_>(f) \in \text{in}_>(I)$, but $\text{in}_>(f)$ must be equal to $\underline{s}_i$ for some i, a contradiction. Thus M is linearly independent in S/I.

Next, if possible, assume that M does not span S/I.
Consider $\{\text{in}_>(f) \mid f \notin$ the span of $M \cup I\}$. By Artinian property, there exists an element $g \notin$ the span of $M \cup I$, with minimal initial term, $\text{in}_>(g) = \underline{m}$, say. Suppose $\underline{m} \in M$. Then $g - \underline{m}$ has an initial term $\underline{m}' < \underline{m}$ and by the minimality of $\underline{m}$, we have $g - \underline{m}$ is in the vector space span of $M \cup I$. Since $\underline{m} \in M$, we have that g is in the vector space span of $M \cup I$, a contradiction of the choice of g. Hence we obtain that $\underline{m} \notin M$. Note that all monomials of S are in $M \cup \text{in}_>(I)$. Since we have shown that $\underline{m} \notin M$, we have $\underline{m} \in \text{in}_>(I)$. Therefore (since $\underline{m}$ is a monomial), there exists $f \in I$ such that $\text{in}_>(f) = \underline{m} = \text{in}_>(g)$. Thus $\text{in}_>(f - g) < \text{in}_>(g)$, and by the minimality of $\text{in}_>(g)$, $f - g$ is in the span of $M \cup I$. Since $f \in I$, this implies g is in the span of $M \cup I$, a contradiction. Therefore, such a g does not exist, and our assumption is wrong. Hence M spans S/I. The result follows. □

Definition 4.1.12. A linear map $\lambda : \mathbb{R}^r \to \mathbb{R}$ is called a *weight function*. It is called *integral* if it comes from a linear map $\mathbb{Z}^r \to \mathbb{Z}$.

Given any weight function λ, we define a partial order $>_\lambda$ on monomials of S:

$$x^{\underline{a}} >_\lambda x^{\underline{b}}, \text{ if } \lambda(\underline{a}) > \lambda(\underline{b}).$$

Note that $>_\lambda$ is only a partial order, since we may have $\lambda(\underline{a}) = \lambda(\underline{b})$ when $\underline{a} \neq \underline{b}$.

A weight function λ is *compatible* with a given monomial order $>$, if $x^{\underline{a}} >_\lambda x^{\underline{b}}$ implies $x^{\underline{a}} > x^{\underline{b}}$ (thus, the order $>$ is a refinement of $>_\lambda$). Let $f \in S, f = \sum a_j m_j$, m_j being monomials. For a weight function λ, we define $\text{in}_\lambda(f) = \sum_i a_i m_i$, where the sum is over $\{i \mid m_i \text{ max under } >_\lambda\}$. The ideal $\text{in}_\lambda(I)$ is defined similarly.

Remark 4.1.13. Suppose λ is a weight function compatible with a monomial order $>$. Then for any monomial m, $\lambda(m) \geq 0$. For, if $\lambda(m) < 0$, then $m <_\lambda 1$, and therefore $m < 1$ by compatibility, a contradiction.

4.2 Gröbner Basis

Definition 4.2.1. Given $\mathcal{G} = \{g_1, \ldots, g_n\} \subseteq S$ and a monomial order $>$ on S, $\mathcal{G}$ is a *Gröbner basis* for the ideal I generated by $\mathcal{G}$, if $\text{in}_>(I)$ is generated by $\{\text{in}_>(g_1), \ldots, \text{in}_>(g_n)\}$.

Remark 4.2.2. If $I = \langle f_1, \ldots, f_n \rangle$, then $\{f_1, \ldots, f_n\}$ is not necessarily a Gröbner basis for I.

Example 4.2.3. Let us illustrate the previous remark. Let $S = K[x, y]$, and $I = \langle xy, x - y \rangle$ for the monomial order derived from $y > x$. Since $x(x - y) \in I$, we have $x^2 \equiv xy (\text{mod } I) \equiv 0 (\text{mod } I)$. If we simply take the initial terms of the generators of I, we have the set $\{xy, y\}$. Clearly, $x^2 \in \text{in}_>(I)$, but $x^2 \notin \langle xy, y \rangle$, thus the initial terms of the generators of I do not form a Gröbner basis for I.

Existence of a Gröbner basis: Let $I = \langle f_1, \ldots, f_n \rangle$. One can adjoin elements of I to the set $\{f_1, \ldots, f_n\}$ until the initial terms generate $\text{in}_>(I)$. (This is possible because S is Noetherian.)

Definition 4.2.4. Let $\mathcal{G} = \{g_1, \ldots, g_n\}$ be a Gröbner basis for I. Then $\mathcal{G}$ is a *reduced* Gröbner basis if

1. The leading coefficient of $\text{in}_>(g_i)$ is 1 for all $1 \leq i \leq n$.
2. For all i, any of the monomials appearing in g_i do not occur in $\{\text{in}_>(g_j) \mid j \neq i\}$.

A reduced Gröbner basis is also a "minimal" Gröbner basis.

Lemma 4.2.5. *Suppose $\{g_1, \ldots, g_n\} \subseteq I$ are such that $\text{in}_>(I)$ is generated by $\{\text{in}_>(g_1), \ldots, \text{in}_>(g_n)\}$. Then $\{g_1, \ldots, g_n\}$ generates I (and thus is a Gröbner basis).*

Proof. Let $J = \langle g_1, \ldots, g_n \rangle \subseteq I$. We claim that $J = I$ (which will complete the proof). If possible, assume that $J \subsetneq I$. Let f be a polynomial in $I \setminus J$ with the smallest initial term among $\{\text{in}_>(h) \mid h \in I \setminus J\}$.

We have $\text{in}_>(J) \subseteq \text{in}_>(I)$. For all i, we also have $\text{in}_>(g_i) \in \text{in}_>(J)$; while, by hypothesis, $\text{in}_>(I) = \langle \text{in}_>(g_i), i = 1, \cdots, n \rangle$. Therefore, $\text{in}_>(I) \subseteq \text{in}_>(J)$, implying $\text{in}_>(I) = \text{in}_>(J)$.

Thus, although $f \notin J$, we have $\text{in}_>(f) \in \text{in}_>(I) = \text{in}_>(J)$. Hence there exists $g \in J$ such that $\text{in}_>(g) = \text{in}_>(f)$. Therefore, $f - g \in I$ has initial term strictly smaller than $\text{in}_>(f)$, implying (by minimality of the choice of F) that $f - g \in J$. Since $g \in J$, we have $f \in J$, a contradiction. Therefore, our assumption is wrong and $J = I$, as required. □

Corollary 4.2.6. *If $J \subseteq I$ such that $\text{in}_>(J) = \text{in}_>(I)$, then $J = I$.*

4.3 Compatible Weight Orders

Given a monomial order $>$, there exists a weight order λ that is compatible with $>$. In fact, given $>$ and a finite set of monomial comparisons $\{m_1 > n_1, \ldots, m_t > n_t\}$, there exists a compatible weight order λ such that $m_i >_\lambda n_i$ for $1 \leq i \leq t$. We will prove this fact, but we first need some preliminary results on convex sets.

Lemma 4.3.1. *Given a monomial order $>$, let $P = \{\underline{a} - \underline{b} \mid x^{\underline{a}} > x^{\underline{b}}\} \subseteq \mathbb{Z}^r$. Then*

1. *If $u \in P$, then $-u \notin P$.*
2. *If $u, v \in P$ and $p, q \in \mathbb{Q}_+$ such that $pu + qv \in \mathbb{Z}^r_+$, then $pu + qv \in P$.*

(In other words, P is the set of positive, integral points of a strictly convex cone in $\mathbb{R}^r$.)

Proof. Part (I) is clear.

To prove part (II), let $u = \underline{a} - \underline{b}$ and $v = \underline{c} - \underline{d}$ so that $x^{\underline{a}} > x^{\underline{b}}$ and $x^{\underline{c}} > x^{\underline{d}}$. Let $pu + qv \in \mathbb{Z}^r_+$. Now take $m = x^{pu+qv}$ and $n = 1$. We have $m > 1$, therefore $pu + qv - \underline{0} \in P$, and the result follows. □

Fix $v_0 \in \mathbb{R}^r$ and $c \in \mathbb{R}$. Then $H = \{u \in \mathbb{R}^r \mid v_0 \cdot u = c\}$ is a hyperplane; and it determines two closed (resp. open) half spaces, namely $\{u \mid u \cdot v_0 \geq c\}$ and $\{u \mid u \cdot v_0 \leq c\}$ (resp. $\{u \mid u \cdot v_0 > c\}$ and $\{u \mid u \cdot v_0 < c\}$).

Lemma 4.3.2. *Let S be a closed, bounded, convex set in $\mathbb{R}^r$. For $p \in \mathbb{R}^r$, either p belongs to S, or there exists a hyperplane H such that $p \in H$ and S is contained in one of the half spaces determined by H.*

Proof. Let $p \notin S$. Define $f : S \to \mathbb{R}$ such that $f(x) = \|x - p\|$. Now S being closed and bounded, is compact, and hence f has a minimum on S. Let $q \in S$ be such that

$$\|q - p\| \leq \|x - p\|, \quad \forall x \in S.$$

Let $n = q - p$. Since $p \notin S$, $n \neq \underline{0}$. Let H be the hyperplane passing through p and orthogonal to n. We will show that H and H_0 have the stated properties.

Let H_0 be the half space $H_0 = \{x \mid x \cdot n \geq p \cdot n\}$. Let $q' \in S$, $q' \neq q$. Then for $0 < t \leq 1$, we have (by the convexity of S)

$$\|q - p\| \leq \|q + t(q' - q) - p\| = \|(q - p) + t(q' - q)\|.$$

We square both sides to get

$$(q - p)^2 \leq (q - p)^2 + 2t(q - p) \cdot (q' - q) + t^2(q' - q)^2$$

(where $(q - p)^2$ denotes $(q - p) \cdot (q - p)$). Canceling terms and dividing by t yields

$$0 \leq 2(q - p) \cdot (q' - q) + t(q' - q)^2.$$

As t goes to zero, we have

$$\begin{aligned} 0 &\leq (q - p) \cdot (q' - q) \\ &= n \cdot q' - n \cdot q \\ &= n \cdot (q' - p) + n \cdot (p - q) \\ &= n \cdot (q' - p) - n \cdot n \end{aligned}$$

Since $n \cdot n > 0$, we have $n \cdot (q' - p) > 0$, and therefore $q' \cdot n > p \cdot n$. We have shown that every $q' \neq q$ in S is contained in H_0. For q, note that $q \cdot n = (p + n) \cdot n = p \cdot n + n \cdot n > p \cdot n$; thus $q \in H_0$, and the result follows. □

Definition 4.3.3. Let S be a convex set in $\mathbb{R}^r$, and p a boundary point of S. A hyperplane H is said to be a *supporting hyperplane of* S at p if $p \in H$ and S is contained in one of the closed half spaces determined by H.

In general, for $f : \mathbb{R}^r \to \mathbb{R}$, the kernel of f is a hyperplane, call it H_f, and any translates of H_f are hyperplanes; i.e., for $a \in \mathbb{R}^r$, $H_f + a$ is a hyperplane. Then H_f defines half spaces as above, $(H_f)_{>0} = \{v \in \mathbb{R}^r \mid f(v) > 0\}$, etc. All of the above holds over $\mathbb{Q}$ also.

We now return to showing the existence of compatible weight orders.

Lemma 4.3.4. *Given a monomial order* $>$ *on* S, *and a finite set of monomial comparisons* $\{m_1 > n_1, \ldots, m_t > n_t\}$, *there exists a compatible weight order* λ *such that* $m_i >_\lambda n_i$ *for* $1 \leq i \leq t$.

Proof. Let $m_i = x^{\underline{a}_i}$ and $n_i = x^{\underline{b}_i}$. Let T be the convex hull of the finite set $\{\underline{a}_i - \underline{b}_i, 1 \leq i \leq t\}$. We have that $0 \notin T$. By Lemma 4.3.2, let H be the hyperplane through 0 such that T lies in the positive half space determined by H. Let $\lambda : \mathbb{Q}^r \to \mathbb{Q}$ be such that the kernel of λ is H. By replacing λ with a suitable integer, we may suppose λ is integral, and we have $m_i >_\lambda n_i$ for $1 \leq i \leq t$. The compatibility of λ in view of the fact that P (as in Lemma 4.3.1) is contained in the positive half space determined by H. □

Theorem 4.3.5. *Let $\mathcal{G} = \{g_1, \ldots, g_n\}$ be a Gröbner basis for I with respect to a monomial order $>$. Then there exists a compatible weight order λ such that*

1. *$in_\lambda(g_i) = in_>(g_i)$ for $1 \leq i \leq n$,*
2. *$in_\lambda(I) = in_>(I)$.*

Proof. Let $g_i = a_i m_i + \sum_j a_{ij} n_{ij}$, where $\text{in}_>(g_i) = a_i m_i$ (and n_{ij} are monomials), and $a_i, a_{ij} \in K$. We begin the proof by constructing λ as in Lemma 4.3.4 for a finite collection

$$\bigcup_{i=1}^{n} \left(\bigcup_j \{m_i > n_{ij}\} \right).$$

Thus we have $m_i >_\lambda n_{ij}$, and therefore $\text{in}_\lambda(g_i) = a_i m_i = \text{in}_>(g_i)$.

It remains to be shown that $\text{in}_\lambda(I) = \text{in}_>(I)$. Since $\mathcal{G}$ is a Gröbner basis, we have that $\{\text{in}_>(g_1), \ldots, \text{in}_>(g_n)\}$ generates $\text{in}_>(I)$, and from above, this set is contained in $\text{in}_\lambda(I)$, therefore $\text{in}_>(I) \subseteq \text{in}_\lambda(I)$.

Let $f \in I$. We claim that $\text{in}_\lambda(f) \in \text{in}_>(I)$, which will complete the proof. Let $f = \sum_i a_i m_i$; we will prove the result by induction on the number of terms in $\text{in}_\lambda(f)$ (recall that since λ is a weight order, $\text{in}_\lambda(f)$ may have multiple terms). If $\text{in}_\lambda(f)$ consists of just one term, say $a_1 m_1$, then $\lambda(m_1) > \lambda(m_i)$, $i > 1$. By compatibility, we have $m_1 > m_i$. Therefore, $\text{in}_\lambda(f) = a_1 m_1 = \text{in}_>(f) \in \text{in}_>(I)$, as desired.

Now let $f = \text{in}_\lambda(f) + \sum a_i m_i$, where any monomial appearing in $\text{in}_\lambda(f)$ is $>_\lambda m_i$ for all i. By compatibility, we have that $\text{in}_>(f)$ is one term in $\text{in}_\lambda(f)$. By hypothesis, $\text{in}_>(f) = m\left(\text{in}_>(g_j)\right)$ for some $1 \leq j \leq n$ and some $m \in S$. Since we are proving by induction and assuming $\text{in}_\lambda(f)$ has more than one term, we have $\text{in}_\lambda(f) \neq m\left(\text{in}_>(g_j)\right)$. Note that $(f - mg_j) \in I$, and $\text{in}_\lambda(f - mg_j)$ has fewer terms than $\text{in}_\lambda(f)$ (since, $m\left(\text{in}_>(g_j)\right)$ is a term in $\text{in}_\lambda(f)$). By induction, $\text{in}_\lambda(f - mg_j) \in \text{in}_>(I)$. But $\text{in}_\lambda(f - mg_j) = \text{in}_\lambda(f) - m\left(\text{in}_>(g_j)\right)$, and therefore $\text{in}_\lambda(f) \in \text{in}_>(I)$. □

Remark 4.3.6. We note that certain weight functions have compatible monomial orders: Let $\lambda : \mathbb{R}^r \to \mathbb{R}$ be an integral weight function. We will call λ *nice* if $\lambda(e_i) \geq 0$ for $1 \leq i \leq r$. Let $\omega = (\lambda(e_1), \ldots, \lambda(e_r))$; and let $>$ be any monomial order. We will now define a monomial order $>_\omega$ with which λ is compatible:

$$x^{\underline{a}} >_\omega x^{\underline{b}} \text{ if } \left\{ \begin{array}{l} \lambda(\underline{a}) > \lambda(\underline{b}), \text{ or} \\ \lambda(\underline{a}) = \lambda(\underline{b}) \text{ and } x^{\underline{a}} > x^{\underline{b}} \end{array} \right\}$$

Clearly $>_\omega$ is a total order on monomials of S with which λ is compatible. It is also clear that $m >_\omega 1$ for any monomial $m \in S$.

It remains to be seen that for $x^{\underline{m}} >_\omega x^{\underline{n}}$, we have $x^{\underline{a}} x^{\underline{m}} >_\omega x^{\underline{a}} x^{\underline{n}}$. Note that if $\lambda(\underline{m}) > \lambda(\underline{n})$, then $\lambda(\underline{m} + \underline{a}) = \lambda(\underline{m}) + \lambda(\underline{a}) > \lambda(\underline{n}) + \text{ł}(\underline{a}) = \text{ł}(\underline{n} + \underline{a})$, and thus $x^{\underline{a}} x^{\underline{m}} >_\omega x^{\underline{a}} x^{\underline{n}}$. On the other hand, if $\text{ł}(\underline{m}) = \text{ł}(\underline{n})$, then we must have $x^{\underline{m}} > x^{\underline{n}}$, and since $>$ is a monomial order, we have $x^{\underline{a}} x^{\underline{m}} > x^{\underline{a}} x^{\underline{n}}$. Because $\text{ł}(\underline{m}) + \text{ł}(\underline{a}) = \text{ł}(\underline{n}) + \text{ł}(\underline{a})$, we have $x^{\underline{a}} x^{\underline{m}} >_\omega x^{\underline{a}} x^{\underline{n}}$. Therefore $>_\omega$ is a monomial order. In the construction of $>_\omega$, we used a monomial order $>$, but we do not require compatibility between λ and $>$.

4.4 Flat Families

Definition 4.4.1. Let R be a ring, and M an R-module. M is *flat*, if for any exact sequence of R-modules $0 \to N_1 \to N_2$, the sequence $0 \to M \otimes_R N_1 \to M \otimes_R N_2$ is also exact.

Remark 4.4.2. If M is a free R-module (i.e., M is isomorphic to a direct sum of copies of R), then M is flat (because $M \otimes_R N = N \oplus \ldots \oplus N$).

Definition 4.4.3. Let $f : X \to Y$ be a morphism of schemes. A sheaf $\mathcal{F}$ of $\mathcal{O}_X$-modules is *flat over* Y *at point* $x \in X$ if the stalk $\mathcal{F}_x$ is a flat $\mathcal{O}_{Y,f(x)}$-module; $\mathcal{F}$ is *flat over* Y if it is flat over Y at every $x \in X$. The morphism f is *flat* if $\mathcal{O}_X$ is flat over Y.

Definition 4.4.4. Given $f : X \to Y$ a morphism of schemes, the collection $\{f^{-1}(y) \mid y \in Y\}$ consisting of fibers of f is called a *family of schemes parametrized by* Y. Such a family is said to be a *flat family* if f is flat.

Example 4.4.5. To give a family of closed subschemes of a scheme X parametrized by a scheme Y is simply to give a closed subscheme Z of $X \times Y$; for the fiber $p_2^{-1}(y)$ under $p_2 : Z \to Y$ gets identified with a closed subscheme of X.

Remark 4.4.6. In general, the fibers of a morphism can vary in a highly discontinuous manner. Flatness prevents this pathological behavior. For instance, if $f : X \to Y$ is flat, the fibers are equidimensional, i.e., the fibers have the same dimension. In addition, if the fibers are projective varieties, then they all have the same Hilbert polynomial (for definition, cf. [21, §1.9]).

Remark 4.4.7. When the parametrizing space Y of a flat family is taken to be $\mathbb{A}^1 = \mathrm{Spec}K[t]$, then the family becomes a 1-parameter family parametrized by K. In this situation, if $u \in K$ is invertible, then the fiber at u is called the *generic fiber*. If $u = 0$, the fiber at u is called the *special fiber*.

Let λ be a "nice" weight function. Let $g \in S$, say $g = (a_1m_1 + \ldots + a_sm_s) + (a_{s+1}m_{s+1} + \ldots + a_vm_v)$ where $\lambda(m_1) = \ldots = \lambda(m_s)$ and $\lambda(m_1) > \lambda(m_j)$ for $j \geq s+1$. Thus $\mathrm{in}_\lambda(g) = \sum_{i=1}^{s} a_im_i$. For simplicity, let us denote $\lambda(m_1)$ by $n(g)$ and $\lambda(m_i)$ by n_i. We define $\tilde{g} \in S[t]$ as:

$$\begin{aligned}\tilde{g} &= t^{n(g)} \sum_{i=1}^{v} a_i t^{-n_i} m_i \\ &= a_1m_1 + \ldots + a_sm_s + \sum_{i=s+1}^{v} a_i t^{n(g)-n_i} m_i.\end{aligned}$$

For any ideal $I \subset S$, let $\tilde{I}$ be the ideal in $S[t]$ generated by $\{\tilde{g} \mid g \in I\}$.

Consider the morphism $f : \operatorname{Spec} S[t]/\tilde{I} \to K$, induced by the canonical map $f^* : K[t] \to S[t]/\tilde{I}$. In the following theorem, we show that f is flat (cf. Definition 4.4.4). Thus we obtain a flat family parametrized by $\mathbb{A}^1$.

Theorem 4.4.8. *As above, let λ be a nice weight function with compatible monomial order $>$. Let $\mathcal{B} = \{m \mid m \notin in_>(I)\}$ where m is a monomial. Then*

1. *$\mathcal{B}$ is a $K[t]$-basis for $S[t]/\tilde{I}$ (and thus $S[t]/\tilde{I}$ is flat over $K[t]$; hence f is flat).*
2. *$S[t]/\tilde{I} \otimes_{K[t]} K[t,t^{-1}] \cong S/I \otimes_K K[t,t^{-1}]$ (thus the generic fiber is S/I, i.e., the fiber over $(t-u)$ for u nonzero in K is S/I).*
3. *$S[t]/\tilde{I} \otimes_{K[t]} K[t]/\langle t\rangle \cong S/in_\lambda(I)$ (thus the special fiber (fiber over zero) is $S/in_\lambda(I)$).*

Proof. We will prove the claims above in the opposite order in which they were stated. Part (III) follows from the fact that

$$S[t]/\left(\tilde{I} + \langle t\rangle\right) = S/\mathrm{in}_\lambda(I).$$

Let $\phi : S[t,t^{-1}] \cong S[t,t^{-1}]$ be the automorphism, $\phi(x_i) = t^{\lambda(x_i)}x_i$. Then $\phi(\tilde{g}) = t^{n(g)}g$. Hence ϕ induces an isomorphism $\tilde{I}S[t,t^{-1}] \cong IS[t,t^{-1}]$ and therefore induces an isomorphism

$$\phi : S[t,t^{-1}]/\tilde{I}S[t,t^{-1}] \cong S[t,t^{-1}]/IS[t,t^{-1}].$$

Part (II) follows.

To prove part (I), we begin by establishing the linear independence of $\mathcal{B}$ in $S[t]/\tilde{I}$. By Theorem 4.1.11, $\mathcal{B}$ is a K-basis for S/I. Hence by base extension, $\mathcal{B}$ is a $K[t,t^{-1}]$-basis for $S[t,t^{-1}]/IS[t,t^{-1}] = S/I\,[t,t^{-1}]$. Now $\phi^{-1}(m) = t^{-\lambda(m)}m$; hence we obtain that $\mathcal{B}$ is a $K[t,t^{-1}]$-basis for $S[t,t^{-1}]/\tilde{I}S[t,t^{-1}]$; in particular, $\mathcal{B}$ is $K[t,t^{-1}]$-linearly independent. Therefore, $\mathcal{B}$ considered as a subset of $S[t]/\tilde{I}$ is $K[t]$-linearly independent. (Note: under the projection $S \to S/I$, $\mathcal{B}$ is mapped injectively; for if $m_1 - m_2 \in I$, then either m_1 or m_2 is in $\mathrm{in}_>(I)$, which is not true if they are members of $\mathcal{B}$.)

We now show that $\mathcal{B}$ generates $S[t]/\tilde{I}$ as a $K[t]$-module. Regarding $\mathcal{B}$ as a subset of $S[t]$, we must show that the $K[t]$-span of $\mathcal{B}$ contains all the monomials modulo $\tilde{I}$. To begin, $m = 1$ is clearly in $\mathcal{B}$. Since the order $>$ is Artinian, we may inductively assume that the result holds for every monomial less than a given monomial m. If m is in $\mathcal{B}$, we have nothing to prove. Let then $m = \mathrm{in}_>(g)$ for some $g \in I$. (Note that $\mathcal{B} \cup \{\mathrm{in}_>(f) \mid f \in I\}$ gives all of the monomials in S.) Let $\tilde{g} = a_1m_1 + \ldots + a_sm_s + \sum_{i=s+1}^{v} t^{n(g)-n_i}m_i$ as in the discussion at the beginning of the section. Then the compatibility of λ with $>$ implies that $\mathrm{in}_>(g) \in \{m_1,\ldots,m_s\}$. Say $m_1 = m = \mathrm{in}_>(g)$. We may suppose $a_1 = 1$. Thus $m - \tilde{g}$ is a $K[t]$-linear combination of monomials that are less than m. We are done by induction (note that $\tilde{g} \in \tilde{I}$ since $g \in I$). □

Remark 4.4.9. We note that in the proof above, the compatibility of λ and $>$ was used only for the proof that $\mathcal{B}$ generates $S[t]/\tilde{I}$.

Theorem 4.4.10. *Given an ideal I of S and a monomial order $>$, there exists a flat family with generic fiber S/I and special fiber $S/in_{>}I$.*

Proof. Given a monomial order $>$ and an ideal I of S, there exists a Gröbner basis for I with respect to $>$. By Theorem 4.3.5, we have a compatible weight order λ such that $\mathrm{in}_{\lambda}(I) = \mathrm{in}_{>}(I)$. The result follows from Theorem 4.4.8. □

Example 4.4.11. Let us consider the conic $xz - y^2 = 0$ in the projective plane $\mathbb{P}^2$. Let $I = \langle xz - y^2 \rangle$ in $K[x, y, z]$.

1. For the lexicographic monomial order induced by the total order $x > y > z$, then $\mathrm{in}_{>}(I) = \langle xz \rangle$, and thus there is a flat family with special fiber $\{x = 0\} \cup \{y = 0\}$.
2. Now we use the reverse lexicographic monomial order, defined as:

$$x^{a_1}y^{a_2}z^{a_3} >_{rev} x^{b_1}y^{b_2}z^{b_3}, \text{ if } \sum a_i > \sum b_i,$$

   or if the degrees are equal then there exists some $1 \leq t \leq 3$ such that $a_t < b_t$ and $a_s = b_s$ for $s > t$. Thus $\mathrm{in}_{>_{rev}}(I) = \langle y^2 \rangle$. Thus there is a flat family with special fiber the double line $y = 0$.

Remark 4.4.12. When a flat family has been established, many geometric properties for the generic fiber may be inferred by knowing them for the special fiber, for example, normality, Cohen–Macaulayness, Gorenstein, etc. Often the geometric study of the given variety X will be reduced to that of a simpler variety X_0 by constructing a flat family with X as the generic fiber and X_0 as the special fiber. In this setting, we say X *degenerates* to X_0; and X_0 *deforms* to X.

Part II
Grassmann and Schubert Varieties

Chapter 5
The Grassmannian and Its Schubert Varieties

In this chapter, we introduce Grassmannian varieties and their Schubert subvarieties; we have also included a brief introduction to flag varieties. We present the Standard Monomial Theory (abbreviated SMT) for Grassmannian and Schubert varieties. As a prime application of SMT, we present a proof of the vanishing of the higher cohomology groups of ample line bundles on the Grassmannian (and their restrictions to Schubert varieties).

5.1 Grassmannian and Flag Varieties

Throughout this chapter, we work over an algebraically closed field K of arbitrary characteristic. Let $G = SL_n(K)$ be the group of $n \times n$ matrices of determinant 1 with entries in K. Let B be the subgroup of "upper triangular" matrices in G, where for $b \in B$, all entries of b below the diagonal are zero.

Definition 5.1.1. Call any subgroup of G conjugate to the subgroup of upper triangular matrices a *Borel* subgroup; i.e., if B' is a subgroup of G, B' is a Borel subgroup if there exists $g \in G$ such that

$$B' = gBg^{-1}.$$

Definition 5.1.2. A closed subgroup of G containing a Borel subgroup is called a *parabolic* subgroup.

Let B be a Borel subgroup of G. For convenience, we may consider B to be the subgroup of "upper triangular" matrices in G, as above. The maximal parabolic subgroups of G that contain this choice of B are of the form P_d, $1 \le d \le n-1$, where

V. Lakshmibai, J. Brown, *The Grassmannian Variety*,
Developments in Mathematics 42, DOI 10.1007/978-1-4939-3082-1_5

$$P_d = \left\{ \begin{bmatrix} * & * \\ 0_{(n-d)\times d} & * \end{bmatrix} \in G \right\},$$

the subgroup of matrices of determinant equal to one with a $(n-d)\times d$ block of entries equal to zero in the lower left corner.

More generally, let $\underline{d} = (d_1, \ldots, d_r)$, where $1 \le d_1 < d_2 < \ldots < d_r \le n-1$ for some r, $1 \le r \le n-1$. We define $P_{\underline{d}}$ to be the intersection of parabolic subgroups

$$P_{\underline{d}} = \bigcap_{i=1}^{r} P_{d_i}.$$

(Note that $P_{\underline{d}}$ is a closed subgroup containing B, and thus is also a parabolic subgroup.)

Definition 5.1.3. A *flag* (or also a *full flag*) in K^n, is a sequence

$$(0) = V_0 \subset V_1 \subset V_2 \subset \cdots \subset V_n = K^n,$$

where V_i is a subspace of K^n of dimension i. For example, we may take V_i to be the span of the first i standard basis vectors $e_1, \cdots, e_i$; this is called *the standard flag*. We denote $\mathcal{F}l_n :=$ { all flags in K^n}.

We have an identification of $\mathcal{F}l_n$ with $GL_n/B_n, B_n$ being the Borel subgroup of GL_n consisting of upper triangular matrices. To see this identification, we first observe that we have a natural transitive action of GL_n on $\mathcal{F}l_n$. Further, under this action, it is seen easily that the stabilizer of the standard flag is precisely B_n. From this, the above-mentioned identification follows. Since SL_n/B is canonically isomorphic to GL_n/B_n, we also have an identification of $\mathcal{F}l_n$ with SL_n/B. More generally, given $\underline{d} = (d_1, \ldots, d_r)$, where $1 \le d_1 < d_2 < \ldots < d_r \le n-1$ for some r, $1 \le r \le n-1$, let $\mathcal{F}l_{n,\underline{d}}$ be the set of all type $\underline{d}$ flags in K^n, i.e.,

$$\mathcal{F}l_{n,\underline{d}} = \{V_{d_1} \subset V_{d_2} \subset \cdots \subset V_{d_r},\ dim\, V_{d_j} = d_j, 1 \le j \le r\}.$$

As above, we have an identification of $\mathcal{F}l_{n,\underline{d}}$ with $SL_n/P_{\underline{d}}$. Thus $\mathcal{F}l_n$, $\mathcal{F}l_{n,\underline{d}}$ acquire projective variety structure, and are respectively called the *full flag variety*, *partial flag variety* (of type $\underline{d}$).

As a special case, when $r = 1$, we have that $\mathcal{F}l_{n,\underline{d}}$ is the celebrated *Grassmannian variety*, consisting of all d-dimensional subspaces of K^n; in the sequel, we shall denote it by $G_{d,n}$. The projective variety structure on $G_{d,n}$ is described in the following section. For any $\underline{d}$, the projective variety structure on $\mathcal{F}l_{n,\underline{d}}$ may be deduced using the embedding

$$\mathcal{F}l_{n,\underline{d}} = \{(V_{d_1}, V_{d_2}, \cdots, V_{d_r}) \in \prod_{i=1}^{r} G_{d_i,n} \mid V_{d_1} \subset V_{d_2} \subset \cdots \subset V_{d_r}\}.$$

The incidence relations $V_{d_{i-1}} \subset V_{d_i}, 2 \le i \le r$ identify $\mathcal{F}l_{n,\underline{d}}$ as a closed subvariety of $\prod_{i=1}^{r} G_{d_i,n}$.

Remark 5.1.4. Before continuing to the variety structure on $G_{d,n}$, let us include the general definition of a Borel subgroup of an algebraic group. Let G be an algebraic group over K. For $g_1, g_2 \in G$, define

$$(g_1, g_2) = g_1 g_2 g_1^{-1} g_2^{-1}.$$

Let G_1 be the *commutator subgroup* of G, i.e., the subgroup generated by all elements of the form (g_1, g_2), $g_1, g_2 \in G$. We will use the notation $G_1 = (G, G)$. For $i \geq 1$, define $G_{i+1} = (G_i, G_i)$; G_i's are closed subgroups of G. The group G is *solvable* if G_n is the trivial subgroup for some n. A maximal connected solvable subgroup of G is called a *Borel* subgroup.

5.2 Projective Variety Structure on $G_{d,n}$

In this section, we shall describe the projective variety structure on $G_{d,n}$.

5.2.1 *Plücker coordinates*

We first define the *Plücker map* $p : G_{d,n} \to \mathbb{P}\left(\bigwedge^d K^n\right)$ as follows. Let $U \in G_{d,n}$ have basis $\{u_1, \ldots, u_d\}$, we define $p(U) := [u_1 \wedge \cdots \wedge u_d]$. To see that this is well defined, suppose we choose a different basis for U: $\{u'_1, \ldots, u'_d\}$. If we let M be the $n \times d$ matrix with columns $u_1, \ldots, u_d$ (each u_i being expressed as a column vector with respect to the standard basis $\{e_i, 1 \leq i \leq n\}$ of K^n), and M' the matrix with columns $u'_1, \ldots, u'_d$, then there exists an invertible $d \times d$ matrix C such that $M = M'C$. Then we have

$$u_1 \wedge \cdots \wedge u_d = \det(C) u'_1 \wedge \cdots \wedge u'_d$$

and thus the map p is well defined.

Let $I_{d,n} = \{\underline{i} = (i_1, \ldots, i_d) \mid 1 \leq i_1 < i_2 < \ldots < i_d \leq n\}$. As above, we shall denote the standard basis for K^n by $\{e_i, 1 \leq i \leq n\}$. We will use the notation $e_{\underline{i}} = e_{i_1} \wedge \cdots \wedge e_{i_d} \in \bigwedge^d K^n$. Thus $\{e_{\underline{i}} \mid \underline{i} \in I_{d,n}\}$ serves as a basis for $\bigwedge^d K^n$. Let $\{p_{\underline{j}} \mid \underline{j} \in I_{d,n}\}$ be the basis of $\left(\bigwedge^d K^n\right)^*$ dual to $\{e_{\underline{i}}\}$; then $\{p_{\underline{j}} \mid \underline{j} \in I_{d,n}\}$ gives a set of projective coordinates for $\mathbb{P}\left(\bigwedge^d K^n\right)$, called the *Plücker coordinates.*

For points $U \in G_{d,n}$, the Plücker coordinates $p_{\underline{j}}(U)$, $\underline{j} \in I_{d,n}$ have a nice description as given below. Let $\{u_1, \ldots, u_d\}$ be a basis for U, and as above, let A be the $n \times d$ matrix with columns $u_1, \ldots, u_d$. Then $p_{\underline{j}}(U)$ is simply the d-minor of A with row indices $j_1, \ldots, j_d$; in the sequel, we shall denote this d-minor by just $\det A_{\underline{j}}$. Thus $p_{\underline{j}}(U) = \det A_{\underline{j}}$.

Theorem 5.2.1. *The Plücker map is injective.*

Proof. Suppose U and U' are two elements of $G_{d,n}$ such that $p(U) = p(U')$. As above, let A and A' be the $n \times d$ matrices corresponding to U and U', respectively. Since $p(U) = p(U')$, there exists $\underline{l} \in I_{d,n}$ such that $p_{\underline{l}}(U)$ and $p_{\underline{l}}(U')$ are both nonzero, and since we are working with projective coordinates, we may assume $p_{\underline{l}}(U) = p_{\underline{l}}(U') = 1$ (and all other Plücker coordinates of U and U' are equal).

As seen above, we may replace A and A' with the matrices $A(A_{\underline{l}})^{-1}$ and $A'(A'_{\underline{l}})^{-1}$, respectively. This replacement allows us to assume that $A_{\underline{l}} = A'_{\underline{l}} = I_d$, the $d \times d$ identity matrix.

Using the notation $A = (a_{ij})$ and $A' = (a'_{ij})$, note that we can isolate any particular entry a_{ij} by taking the d-minor of A with row indices $\underline{l}$ but replacing l_j with the entry i; i.e.,

$$a_{ij} = \det A_{l_1,\ldots,l_{j-1},i,l_{j+1},\ldots,l_d} = p_{l_1,\ldots,l_{j-1},i,l_{j+1},\ldots,l_d}(U).$$

Similarly,

$$a'_{ij} = \det A'_{l_1,\ldots,l_{j-1},i,l_{j+1},\ldots,l_d} = p_{l_1,\ldots,l_{j-1},i,l_{j+1},\ldots,l_d}(U').$$

Since $p(U) = p(U')$, we obtain that $a_{ij} = a'_{ij}$, which implies $A = A'$, and thus $U = U'$. □

5.2.2 *Plücker Relations*

In this subsection, we will show that $G_{d,n}$ is the zero set of a system of quadratic polynomials in $\{p_{\underline{i}} \mid \underline{i} \in I_{d,n}\}$.

Definition 5.2.2. Let $\underline{i} \in I_{d-1,n}$ and $\underline{j} \in I_{d+1,n}$, and consider the following quadratic polynomial in the Plücker coordinates, called a Plücker polynomial:

$$\sum_{h=1}^{d+1} (-1)^h p_{i_1,\ldots,i_{d-1},j_h} p_{j_1,\ldots,\widehat{j_h},\ldots,j_{d+1}}.$$

In the equation above, if $\underline{a}$ has a repeated entry, $p_{\underline{a}}(A)$ is understood to be zero for any $n \times d$ matrix A. We shall denote the above system of polynomials by $\mathcal{P}_{d,n}$

For example, in $G_{2,4}$, if we let $\underline{i} = \{1\}$ and $\underline{j} = \{2,3,4\}$, then we have the polynomial

$$-p_{1,2}p_{3,4} + p_{1,3}p_{2,4} - p_{1,4}p_{2,3}.$$

In fact, $\mathcal{P}_{2,4}$ consists of only the polynomial above (up to sign). In general, $\mathcal{P}_{d,n}$ contains many polynomials, as the choices for $\underline{i}$ and $\underline{j}$ increase.

Theorem 5.2.3. *The image of $G_{d,n}$ in $\mathbb{P}\left(\bigwedge^d K^n\right)$ is precisely the zero set of the system of polynomials $\mathcal{P}_{d,n}$.*

Proof. First, we will show that one obtains zero when evaluating any polynomial in $\mathcal{P}_{d,n}$ at U for any $U \in G_{d,n}$. Let $A = (a_{ij})$ be the $n \times d$ matrix representing U. We will expand the d-minor $A_{i_1,\dots,i_{d-1},j_h}$ along the last row j_h. This gives

$$\sum_{h=1}^{d+1}(-1)^h p_{i_1,\dots,i_{d-1},j_h}(U)p_{j_1,\dots,\widehat{j_h},\dots,j_{d+1}}(U)$$

$$=\sum_{h=1}^{d+1}(-1)^h\sum_{m=1}^{d}(-1)^{d+m}a_{j_h,m}\det(A^{\hat{m}}_{i_1,\dots,i_{d-1}})\det(A_{j_1,\dots,\widehat{j_h},\dots,j_{d+1}})$$

where $A^{\hat{m}}_{i_1,\dots,i_{d-1}}$ represents deleting the mth column yielding a $(d-1)$-minor, and after some rearranging this is equal to

$$\sum_{m=1}^{d}(-1)^{d+m}\det(A^{\hat{m}}_{i_1,\dots,i_{d-1}})\left(\sum_{h=1}^{d+1}(-1)^h a_{j_h,m}\det(A_{j_1,\dots,\widehat{j_h},\dots,j_{d+1}})\right).$$

The second summation above represents the determinant of the following $(d+1)\times(d+1)$ matrix (up to sign)

$$\begin{bmatrix} a_{j_1,m} & a_{j_1,1} & a_{j_1,2} & \cdots & a_{j_1,d} \\ a_{j_2,m} & a_{j_2,1} & a_{j_2,2} & \cdots & a_{j_2,d} \\ \vdots & \vdots & \vdots & \ddots & \vdots \\ a_{j_{d+1},m} & a_{j_{d+1},1} & a_{j_{d+1},2} & \cdots & a_{j_{d+1},d} \end{bmatrix}.$$

Note that since m is between 1 and d, this matrix has a repeated column, and thus its determinant is zero. Therefore

$$\sum_{h=1}^{d+1}(-1)^h p_{i_1,\dots,i_{d-1},j_h}(U)p_{j_1,\dots,\widehat{j_h},\dots,j_{d+1}}(U)=0.$$

Now we show that $G_{d,n}$ contains the zero set of $\mathcal{P}_{d,n}$. Let $q = (q_{\underline{i}}) \in \mathbb{P}\left(\bigwedge^d K^n\right)$ be a zero of $\mathcal{P}_{d,n}$. Since we are working in projective coordinates, we may assume that there exists $\underline{l} \in I_{d,n}$ such that $q_{\underline{l}} = 1$. Define the $n \times d$ matrix A such that $a_{ik} = q_{l_1,\dots,l_{k-1},i,l_{k+1},\dots,l_d}$. This implies $A_{\underline{l}}$ is the $d \times d$ identity matrix, and thus A has rank d and represents an element of $G_{d,n}$. We shall now prove that $p_{\underline{j}}(A) = q_{\underline{j}}$ for all $\underline{j} \in I_{d,n}$, from which the result will follow. We prove this assertion by decreasing induction on the cardinality of $\underline{j} \cap \underline{l}$. To begin, if $\underline{j} = \underline{l}$, then $p_{\underline{j}}(A) = \det(A_{\underline{l}}) = 1 = q_{\underline{l}}$. If $\underline{j} = (l_1,\dots,l_{k-1},i,l_{k+1},\dots,l_d)$, then $p_{\underline{j}}(A) = a_{i,k} = q_{\underline{j}}$, from our definition of A above, and thus the result follows for $\#\{\underline{j} \cap \underline{l}\} \geq d-1$.

Now let $\underline{j} \in I_{d,n}$; we invoke the equation of 5.2.2 using the tuples $(l_1, \ldots, l_{d-1})$ and $(l_d, j_1, \ldots, j_d)$ (we assume without loss of generality that $l_d \notin \{j_1, \ldots, j_d\}$). Since q is in the zero set, we have

$$q_{\underline{l}} q_{\underline{j}} \pm \sum q_{\underline{l}'} q_{\underline{j}'} = 0$$

where $\#\{\underline{l} \cap \underline{j}'\} > \#\{\underline{l} \cap \underline{j}\}$ because some entry in $(j_1, \ldots, j_d)$ not in $\underline{l}$ has been replaced by l_d in $\underline{j}'$.

Therefore, by the induction hypothesis, $q_{\underline{j}'} = p_{\underline{j}'}(A)$. Since $\underline{l}'$ and $\underline{l}$ differ by only one entry, we also have $q_{\underline{l}'} = p_{\underline{l}'}(A)$.

We have already shown that A is in the zero set of $\mathcal{P}_{d,n}$, thus

$$p_{\underline{l}}(A) p_{\underline{j}}(A) \pm \sum p_{\underline{l}'}(A) p_{\underline{j}'}(A) = 0.$$

Using all of the above, we have that

$$p_{\underline{l}}(A) p_{\underline{j}}(A) = q_{\underline{l}} q_{\underline{j}},$$

and because $p_{\underline{l}}(A) = 1 = q_{\underline{l}}$, we have that $p_{\underline{j}}(A) = q_{\underline{j}}$, and the result follows. □

Remark 5.2.4. Let $S = K[p_{\underline{i}} \mid \underline{i} \in I_{d,n}]$, and I be the ideal in S generated by the Plücker polynomials. Let $J = \mathcal{I}(G_{d,n}) \subset S$, the defining ideal of $G_{d,n}$. From above, we have that $I \subseteq J$. In §5.4, we will show that in fact $I = J$.

5.2.3 Plücker coordinates as T-weight vectors

Let $T \subset G = SL_n$ be the maximal torus of diagonal matrices. Given $t \in T$, we will write $t = (t_1, \ldots, t_n)$, where t_i is the ith diagonal entry of t. We have $T = (K^*)^{n-1}$, and is a torus (the dimension is $n-1$ because $t_n = (t_1 \cdots t_{n-1})^{-1}$).

Let $X(T)$ be the *character group* of T, namely, the group of all algebraic group homomorphisms $\chi : T \to K^*$. (Here, K^* is being viewed as the multiplicative group of the field K.) Define $\varepsilon_i \in X(T)$ such that $\varepsilon_i(t_1, \ldots, t_n) = t_i$. Note that because $t_n = \frac{1}{t_1 \cdots t_{n-1}}$, we have $\varepsilon_n = -(\varepsilon_1 + \ldots + \varepsilon_{n-1})$, the group $X(T)$ being written additively. We have that $\{\varepsilon_i, 1 \leq i \leq n-1\}$ is a $\mathbb{Z}$-basis for $X(T)$.

Consider $e_i \in K^n$. For the natural action of T on K^n (induced by the natural action of G on K^n), we have, $t \cdot e_i = t_i e_i$, $t \in T$. We call e_i, a *T-weight vector* of weight ε_i. More generally, given a T-module V, a vector $v \in V$ such that $t \cdot v = \chi(t) v$ for all $t \in T$ and some $\chi \in X(T)$ is called a *T-weight vector* of weight χ. We have that the element $e_{\underline{i}} \in \bigwedge^d K^n$ is a T-weight vector of weight $\varepsilon_{i_1} + \varepsilon_{i_2} + \ldots + \varepsilon_{i_d}$.

Now we consider the Plücker coordinate $p_{\underline{j}}$. Let us denote the integers $\{1,\ldots,n\}\setminus\{j_1,\ldots,j_d\}$ by $j_{d+1},\ldots,j_n$ (arranged in ascending order). Under the T-equivariant isomorphism

$$\bigwedge^{n-d} K^n \xrightarrow{\sim} \left(\bigwedge^{d} K^n\right)^*$$

$$f \longmapsto f(v) = v \wedge f,\ v \in \bigwedge^{d} K^n$$

$e_{j_{d+1}} \wedge \ldots \wedge e_{j_n}$ is identified with $p_{\underline{j}}$; hence, $p_{\underline{j}}$ is a T-weight vector of weight

$$\varepsilon_{j_{d+1}} + \ldots + \varepsilon_{j_n} = -(\varepsilon_{j_1} + \ldots + \varepsilon_{j_d}).$$

Thus the weight of $p_{\underline{j}}$ is the negative of the weight of $e_{\underline{j}}$.

5.3 Schubert Varieties

Schubert varieties are certain closed subvarieties of the Grassmannian variety, and are indexed by $I_{d,n}$. First, we observe that for the natural action of G on $\mathbb{P}\left(\bigwedge^d K^n\right)$ (induced from the natural action of G on K^n), we have that $G_{d,n}$ is G-stable; in fact, it is easily seen that the action on $G_{d,n}$ is transitive and we have an identification of $G_{d,n}$ with G/P_d. Further, we have that the T-fixed points in $G_{d,n}$ are precisely $e_{\underline{i}} (= [e_{i_1} \wedge \ldots \wedge e_{i_d}])$, $\underline{i} \in I_{d,n}$

Definition 5.3.1. Let $\underline{i} \in I_{d,n}$. The Schubert variety $X_{\underline{i}}$ is the Zariski closure of the B-orbit $B \cdot [e_{i_1} \wedge \ldots \wedge e_{i_d}]$ in $G_{d,n}$, with the canonical reduced structure. We will also use the notation $X(\underline{i})$. The B-orbit $B \cdot [e_{\underline{i}}]$ is called the *Schubert cell*, and is denoted $C_{\underline{i}}$. It is a dense open subset of $X_{\underline{i}}$. Thus Schubert varieties are the closures of B-orbits through the T-fixed points in $G_{d,n}$.

Remark 5.3.2. $G_{d,n} = X_{\underline{i}}$, for $\underline{i} = (n+1-d,\ n+2-d,\ldots,n)$, and the point (corresponding to the coset B) is the Schubert variety $X_{\underline{i}}$, $\underline{i} = (1,2,\ldots,d)$.

Partial order on $I_{d,n}$: Given $\underline{i}, \underline{j}$, we define

$$\underline{i} \geq \underline{j} \text{ if and only if } i_1 \geq j_1,\ i_2 \geq j_2, \ldots, i_d \geq j_d.$$

Example 5.3.3. The following lattice represents the partial order on $I_{3,6}$, where a line segment from $\underline{i}$ down to $\underline{j}$ represents $\underline{i} \geq \underline{j}$.

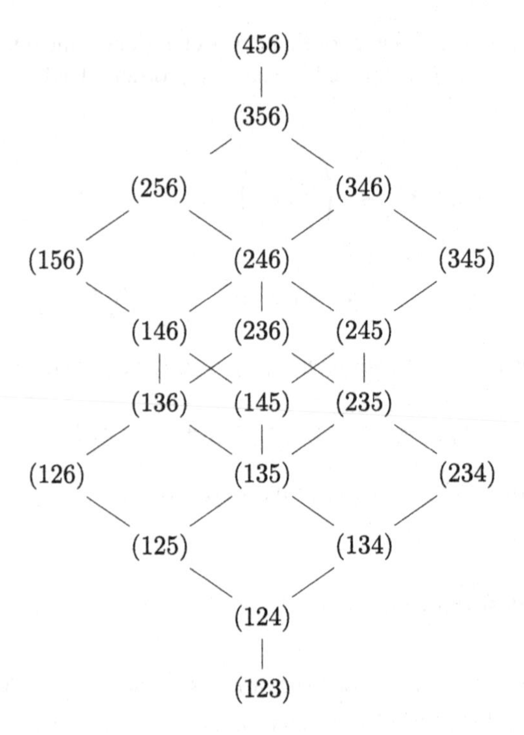

Bruhat decomposition:

$$X_{\underline{i}} = \bigcup_{\underline{j} \leq \underline{i}} C_{\underline{j}}, \quad G_{d,n} = \bigcup_{\underline{i} \in I_{d,n}} C_{\underline{i}}.$$

The classical definition of a Schubert variety in $G_{d,n}$ is as follows. Let V_r be the subspace of K^n spanned by $\{e_1, \ldots, e_r\}$ for $1 \leq r \leq n$. Let $\underline{i} \in I_{d,n}$, and define

$$Y_{\underline{i}} = \{U \in G_{d,n} \mid \dim(U \cap V_{i_t}) \geq t, \ \forall t\}.$$

We shall now see that these two definitions of Schubert varieties ($Y_{\underline{i}}$ and $X_{\underline{i}}$) are equivalent.

First, we observe that $Y_{\underline{i}}$ is B-stable; for, consider $U \in Y_{\underline{i}}$. Then, for $b \in B$, we have, $b \cdot (U \cap V_{i_t}) = b \cdot U \cap V_{i_t}$, since clearly, $b \cdot V_{i_t} = V_{i_t}$. Thus, dim $(b \cdot U \cap V_{i_t}) \geq t, \ \forall t$ (note that dim $(b \cdot (U \cap V_{i_t}))$ = dim $(U \cap V_{i_t})$), and the B-stability of $Y_{\underline{i}}$ follows.

Next, denoting by $U_{\underline{j}}$ the span of $\{e_{j_1}, \ldots, e_{j_d}\}$, we have $\dim(U_{\underline{j}} \cap V_{i_t}) \geq t$ for all $1 \leq t \leq d$ if and only if $\underline{j} \leq \underline{i}$. Together with the B-stability of $Y_{\underline{i}}$, this implies that

$$Y_{\underline{i}} = \bigcup_{\underline{j} \leq \underline{i}} C_{\underline{j}}.$$

In view of Bruhat decomposition, we conclude that $X_{\underline{i}} = Y_{\underline{i}}$.

Remark 5.3.4. For $\underline{i}, \underline{j} \in I_{d,n}$, we have the following:

1. $[e_{j_1} \wedge \cdots \wedge e_{j_d}] \in X_{\underline{i}}$ if and only if $\underline{j} \leq \underline{i}$.
2. $X_{\underline{j}} \subseteq X_{\underline{i}}$ if and only if $\underline{j} \leq \underline{i}$.
3. $p_{\underline{j}}|_{X_{\underline{i}}} \neq 0$ if and only if $\underline{j} \leq \underline{i}$.

The above follow from the facts that $p_{\underline{j}}(e_{\underline{i}}) = \delta_{\underline{i},\underline{j}}$, and that $[e_{\underline{j}}] \in X_{\underline{i}}$ if and only if $\underline{j} \leq \underline{i}$. Thus $X_{\underline{i}}$ consists of all points $P \in G_{d,n}$ such that $p_{\underline{j}}(P) = 0, \underline{j} \not\leq \underline{i}$. Therefore $X_{\underline{i}}$ sits in a smaller projective space.

The following is a result of the above remark combined with Theorem 5.2.3.

Corollary 5.3.5. *Let $\underline{i} \in I_{d,n}$. The image of $X_{\underline{i}}$ in $\mathbb{P}\left(\bigwedge^d K^n\right)$ is precisely the zero set of the set of polynomials given by Definition 5.2.2 in union with the set $\{p_{\underline{j}} \mid \underline{j} \not\leq \underline{i}\}$.*

Example 5.3.6. We have $G_{2,4} \hookrightarrow \mathbb{P}\left(\bigwedge^2 K^4\right) \cong \mathbb{P}^5$, defined by $p_{2,3} \cdot p_{1,4} = p_{2,4} \cdot p_{1,3} - p_{3,4} \cdot p_{1,2}$. Then

$$X_{(2,4)} = G_{2,4} \cap \{p_{3,4} = 0\}.$$

The example above holds in the general case also: Let $\underline{i} = (n-d, n-d+2, n-d+3, \ldots, n)$ and $w_0 = (n-d+1, n-d+2, \ldots, n)$. Then $X_{\underline{i}} = G_{d,n} \cap \{p_{w_0} = 0\}$. In fact, $X_{\underline{i}}$ is the unique Schubert variety of codimension one in $G_{d,n}$ (we will see how to compute dimension in the following discussion). By Bruhat decomposition, we have $G_{d,n} = C_{w_0} \cup X_{\underline{i}}$.

Next, we will find the dimension of X_w for all $w \in I_{d,n}$.

5.3.1 *Dimension of X_w*

Let U be the unipotent part of B (U consists of upper triangular matrices in G with all diagonal entries equal to 1). Then B is the semidirect product of U and T, where T is the maximal torus in B as defined in §5.2.3.

Let $w = (i_1, \cdots, i_d)$, and let $e_w = e_{i_1} \wedge \ldots \wedge e_{i_d}$ as in §5.2.1. Now the dimension of X_w equals that of the B-orbit $B \cdot [e_w]$. The B-orbit $B \cdot [e_w]$ equals the U-orbit $U \cdot [e_w]$ (since, B is the semidirect product of U and T and $[e_w]$ is fixed by T). To compute the dimension of $U \cdot [e_w]$, we shall first determine $\mathrm{stab}_U[e_w]$, the stabilizer of $[e_w]$ in U.

Let $A \in U$, say, $A = Id + \sum_{1 \leq i < j \leq n} c_{ij} E_{ij}$, $c_{ij} \in K$, E_{ij} being the elementary matrix with 1 at the $(i,j)^{th}$ place, and 0's elsewhere, and Id being the identity matrix. Denote $I = \{i_1, \cdots, i_d\}$. Using the fact that for $1 \leq l \leq n$

$$E_{ij} \cdot e_\ell = \begin{cases} e_i, & \text{if } \ell = j \\ 0, & \text{if } \ell \neq j \end{cases}$$

we have the following:

(i) if $j \notin I$, then $E_{ij} \cdot e_\ell = 0$, for all $\ell \in I$,
(ii) if $j \in I$, then $E_{ij} \cdot e_\ell$ is in the span of $\{e_{i_1}, \cdots, e_{i_d}\} \Longleftrightarrow i \in I$.

Hence we obtain $U^w := \mathrm{stab}_U[e_w] =$

$$\{Id + \sum_{1 \le i < j \le n} c_{ij}E_{ij} \in U \mid c_{ij} = 0, \text{ for all } (i,j) \text{ such that } j \in I, \; i \notin I\}.$$

Let us denote the one-dimensional unipotent subgroup of U consisting of $\{Id + c_{ij}E_{ij}, c_{ij} \in K\}$ by U_{ij}. Then taking a total order on the set $\{U_{ij} \mid 1 \le i < j \le n\}$, and writing the set as $\{U_1, \cdots, U_N\}, N = \binom{n}{2}$, we have that as varieties, $U \cong \prod_{1 \le \ell \le N} U_\ell$. From this, it follows that we obtain an identification of the orbit $U \cdot [e_w]$ with the subset U_w of U:

$$U_w = \{Id + \sum_{1 \le i < j \le n} c_{ij}E_{ij} \in U \mid c_{ij} = 0, \text{ for either } j \notin I, \text{ or both } i,j \in I\}.$$

It follows that

$$\dim (U \cdot [e_w]) = \#\{(i,j), 1 \le i < j \le n \,|\, j \in I, i \notin I\}.$$

Thus for $j = i_1$, there are $i_1 - 1$ choices for i. For $j = i_2$, there are $i_2 - 2$ choices for i (since i_1 and i_2 are not allowed). Corresponding to $j = i_t, 1 \le t \le d$, there are $i_t - t$ choices for i. Hence we obtain the following theorem:

Theorem 5.3.7. *Given $w \in I_{d,n}$, we have*

$$\dim X(w) (= \dim U \cdot [e_w]) = \sum_{1 \le t \le d} i_t - t.$$

Corollary 5.3.8. *For the Grassmannian variety,*

$$\dim G_{d,n} = d(n-d).$$

Remark 5.3.9. Let C_w be the open orbit $B \cdot [e_w]$. In the preceding discussion, we have identified C_w with the subset U_w of U as defined above (U_w is in fact a subgroup). Clearly, U_w is also an affine space, and we have an identification of C_w with $\mathbb{A}^{\dim X_w}$.

5.3.2 *Integral Schemes*

A Schubert variety, being a B-orbit closure (with the canonical reduced scheme structure) is an integral scheme. In this section, we give an explicit realization of a

Schubert scheme as an integral scheme. We first consider the case of the Schubert scheme $G_{d,n}$. Let $V = K^n$. Consider

$$\pi : V^{\oplus d} \to \wedge^d V, (v_1, \cdots, v_d) \mapsto v_1 \wedge \cdots \wedge v_d.$$

Then im π is precisely $\widehat{G_{d,n}}(= Spec\, R)$, the cone over $G_{d,n}$ (here, R denotes the homogeneous coordinate ring of $G_{d,n}$ for the Plücker embedding). Thus $\widehat{G_{d,n}}$ is precisely the cone of decomposable vectors in $\wedge^d V$ (a vector in $\wedge^d V$ is decomposable if it can be expressed as the wedge-product of d elements in V). We have that the coordinate rings of the affine spaces $V^{\oplus d}, \wedge^d V$ are the polynomial algebras $K[x_{ij}, 1 \le i \le n, 1 \le j \le d]$, $K[x_{\underline{i}}, \underline{i} \in I_{d,n}]$, respectively. The comorphism

$$\pi^* : K[x_{\underline{i}}, \underline{i} \in I_{d,n}] \to K[x_{ij}, 1 \le i \le n, 1 \le j \le d],$$
$$x_{\underline{i}} \mapsto |A_{\underline{i}}|,$$

$A_{\underline{i}}$ being the $d \times d$ submatrix of $(x_{ij}), 1 \le i \le n, 1 \le j \le d$ with row indices given by $\underline{i}$, identifies R with the subalgebra of $K[x_{ij}, 1 \le i \le n, 1 \le j \le d]$, generated by $\{|A_{\underline{i}}|, \underline{i} \in I_{d,n}\}$. Thus R is an integral domain, and it follows that $G_{d,n}$ is an integral scheme.

Regarding a Schubert variety $X_{\underline{i}}$ in $G_{d,n}$, we have that $X_{\underline{i}}$ is obtained by intersecting $G_{d,n}$ by the hyperplanes $p_{\underline{j}} = 0, \underline{j} \not\le \underline{i}$. From this it follows that $X_{\underline{i}}$ is an integral scheme

5.4 Standard Monomials

Recall from §5.2.1 that for each $\underline{i} \in I_{d,n}$ we have the Plücker coordinate $p_{\underline{i}}$. The Grassmannian variety $G_{d,n}$ is the zero set of the Plücker polynomials in $\mathcal{P}_{d,n}$. Here we give a more general (but equivalent) form of these Plücker polynomials. Let $\underline{i}, \underline{j} \in I_{d,n}$ be such that $\underline{i} \not\ge \underline{j}$; let $r, 1 \le r \le d$ be such that $i_t \ge j_t, 1 \le t \le r-1$ and $i_r < j_r$. Then the "more general" Plücker relations are given by

$$\sum_{\sigma} \text{sign}(\sigma) p_{\sigma(\underline{i})} p_{\sigma(\underline{j})}, \tag{†}$$

where σ runs over all permutations of the set $\{i_1, \ldots, i_r, j_r, \ldots, j_d\}$, such that $\sigma(i_1) < \sigma(i_2) < \ldots < \sigma(i_r); \sigma(j_r) < \ldots < \sigma(j_d)$, and

$$\sigma(\underline{i}) = (\sigma(i_1), \ldots, \sigma(i_r), i_{r+1}, \ldots, i_d),$$

arranged in ascending order (similarly for $\sigma(\underline{j})$). (Also, if $\sigma(\underline{i})$ has repeated entries, $p_{\sigma(\underline{i})}$ is considered zero.) The relations given by $\mathcal{P}_{d,n}$ and those given in (†) above are shown to be equivalent in [89, §3.1].

Remark 5.4.1. Suppose $\underline{i}$ and $\underline{j}$ (in $I_{d,n}$) are not comparable; i.e., there exists some r such that $i_1 \geq j_1, \ldots, i_{r-1} \geq j_{r-1}$, but $i_r < j_r$. Note that if we form a Plücker relation as in (†) using such a choice of $\underline{i}, \underline{j}$, and r, we have that for $\sigma \neq id, \sigma(\underline{i}) > \underline{i}$ (since some $i \in \{i_1, \ldots, i_r\}$ has been replaced by a $j \in \{j_r, \ldots, j_d\}$ and $i_1 < \ldots < i_r < j_r < \ldots < j_d$). Similarly, for $\sigma \neq id, \sigma(\underline{j}) < \underline{j}$.

Example 5.4.2. In $I_{3,6}$, let $\underline{i} = (2,3,4)$ and $\underline{j} = (1,4,5)$, $r = 2$. Then (†) becomes

$$p_{(2,3,4)}p_{(1,4,5)} = p_{(2,4,5)}p_{(1,3,4)} - p_{(3,4,5)}p_{(1,2,4)}.$$

In the sequel, we will also use Greek letters to denote elements of $I_{d,n}$.

Definition 5.4.3. Let $\tau_1, \ldots, \tau_m \in I_{d,n}$. Then $p_{\tau_1} \cdots p_{\tau_m}$ is a *standard monomial* if $\tau_1 \geq \ldots \geq \tau_m$. The same monomial is *standard on the Schubert variety* $X_{\underline{i}}$ if in addition $\underline{i} \geq \tau_1$.

5.4.1 Generation by Standard Monomials

In the rest of §5.4, we show that standard monomials on X_τ form a (vector space) basis for $K[X_\tau]$, the homogeneous coordinate ring of X_τ (for the Plücker embedding). Recall from Remark 5.2.4, we have not yet shown that the ideal generated by the Plücker polynomials is in fact $\mathcal{I}(G_{d,n})$, the defining ideal of the Grassmannian variety (a similar remark for the Schubert subvarieties). For any $\underline{i}, \underline{j} \in I_{d,n}$ and $1 \leq r \leq d$, let $f_{\underline{i},\underline{j},r} = p_{\underline{i}}p_{\underline{j}} - \sum_\sigma \text{sign}(\sigma) p_{\sigma(\underline{i})} p_{\sigma(\underline{j})}$ as in (†) above. Let $S = K[p_{\underline{i}}, \underline{i} \in I_{d,n}]$, and $\tau \in I_{d,n}$. Let $I(\tau)$ be the ideal generated by all possible $f_{\underline{i},\underline{j},r}$, and $\{p_\phi, \phi \not\leq \tau\}$; note that for $\tau = w_0$, the (unique) maximal element of $I_{d,n}$, $I(w_0)$ is the ideal generated by all possible $f_{\underline{i},\underline{j},r}$. In the sequel, we shall denote the ideal $I(w_0)$ by just I. Let $J(\tau) = \mathcal{I}(X_\tau)$, and $J = \mathcal{I}(G_{d,n})$. We shall show that in fact $I(\tau) = J(\tau)$ (in particular $I = J$).

We begin by showing that the set of standard monomials on X_τ is a generating set for $S/I(\tau), \tau \in I_{d,n}$. Clearly, it suffices to show that standard monomials form a generating set for S/I (in view of the natural surjection $S/I \to S/I(\tau)$).

Proposition 5.4.4. *With notation as above, we have that in S/I, any nonstandard monomial is a linear combination of standard monomials.*

Proof. Let $F = p_{\tau_1} \cdots p_{\tau_m}$ be a monomial. Choose $N > d(n-d)$. For $\underline{i} = (i_1, \ldots, i_d) \in I_{d,n}$, let us denote $l(\underline{i}) := \sum_{t=1}^{d} (i_t - t)$. Note that $l(\underline{i}) < N$. Now we define

$$N_F = l(\tau_1)N^{m-1} + l(\tau_2)N^{m-2} + \ldots + l(\tau_m),$$

so that $(l(\tau_1), \ldots, l(\tau_m))$ is the N-ary presentation for N_F. We will show that F is generated by standard monomials by decreasing induction on N_F.

To begin, suppose F has the maximum possible N_F, i.e., $F = p_\theta^m$ for $\theta = (n-d+1,\ldots,n)$. Then clearly F is standard, and the result follows.

Now suppose F is some nonstandard monomial. Thus there exists $1 \le r \le m-1$ such that $\tau_1 \ge \ldots \ge \tau_r \not\ge \tau_{r+1}$. Using Remark 5.4.1, $p_{\tau_r}p_{\tau_{r+1}} = \sum \pm p_\alpha p_\beta$, where $\alpha > \tau_r$ for each α that appears in the summation. Thus $l(\alpha) > l(\tau_r)$, and $F = \sum \pm F_i$, where $N_{F_i} > N_F$ for all i. The result follows by induction. □

As an immediate consequence, we have, using the canonical surjection $\eta_\tau : S/I(\tau) \to S/J(\tau)$, the following:

Corollary 5.4.5. *For $\tau \in I_{d,n}$, the set of standard monomials on X_τ is a generating set for $R(\tau)(= S/J(\tau))$, the homogeneous coordinate ring of X_τ.*

5.4.2 Linear Independence of Standard Monomials

Proposition 5.4.6. *With notation as above, we have that the set of monomials standard on X_τ is linearly independent in $S/J(\tau)$.*

Proof. Since $S/J(\tau)$ is a graded ring, it is enough to show that degree m-monomials standard on X_τ are linearly independent in $S/J(\tau)$, for $m \in \mathbb{N}$. We proceed by induction on $\dim X_\tau$. If $\dim X_\tau = 0$, then $\tau = (1,\ldots,d)$, and p_τ^m is the only standard monomial of degree m on X_τ, so the result follows.

Now let $\dim X_\tau > 0$. Let $\{F_i\}$ be standard monomials on X_τ of degree m. Assume that

$$\sum_{i=1}^{r} a_i F_i = 0, \qquad (*)$$

is a nontrivial linear relation in $S/J(\tau)$, where $a_i \in K$ and $a_i \neq 0$. For each $1 \le i \le r$, let

$$F_i = p_{\tau_{i1}} \cdots p_{\tau_{im}}, \quad \tau_{i1} \ge \ldots \ge \tau_{im}.$$

From the set $\{\tau_{11}, \tau_{21}, \ldots, \tau_{r1}\}$, we choose a minimal element, and call this element w. For simplicity, let us suppose that $w = \tau_{11}$.

Suppose that $w < \tau$. For $i \ge 2$, we have that in $S/J(w)$, F_i is either 0 (namely, if $\tau_{i1} \not\le w$) or is a standard monomial on $X(w)$ (namely, if $\tau_{i1} \le w$ in which case τ_{i1} in fact equals w, by our choice of w). Hence, (*) gives rise to a nontrivial standard sum in $S/J(w)$ being zero (note that when (*) is considered as a relation in $S/J(w)$, the first term in the left hand side of (*) is nonzero). But this is not possible (by induction on dimension). Hence we conclude that $w = \tau$, which in turn implies that $\tau_{i1} = \tau$, $\forall i$. Thus, F_i is divisible by p_τ for all i. Since X_τ is an integral scheme (cf. the end of §5.3.2), we may cancel out p_τ from this equation, leaving a dependence relation of standard monomials in $S/J(\tau)$ of degree $m-1$. By induction on m, all coefficients must be zero, a contradiction. (The case $m = 1$ is clear.) Hence our assumption is wrong and the result follows. □

Proposition 5.4.7. *Let $S = K[p_{\underline{i}}, \underline{i} \in I_{d,n}]$, $J = \mathcal{I}(G_{d,n})$ and $J(\tau) = \mathcal{I}(X_\tau)$ for $\tau \in I_{d,n}$. Then*

1. *The standard monomials on X_τ form a basis for $R(\tau)(= S/J(\tau))$ and*
2. *The canonical surjection $\eta_\tau : S/I(\tau) \to S/J(\tau)$ is in fact an isomorphism.*

Proof. Part (I) follows from the above Proposition and Corollary 5.4.5. Regarding Part (II), first we have that standard monomials on X_τ generate $S/I(\tau)$ (cf. Proposition 5.4.4). Next, the set of standard monomials on X_τ is linearly independent in $S/I(\tau)$, since it is linearly independent in $S/J(\tau)$ (cf. Proposition above). Thus, standard monomials on X_τ form a basis for both $S/I(\tau), S/J(\tau)$; hence the isomorphism as asserted in (2). □

In particular, we have the following result.

Theorem 5.4.8. *The ideal $I(\tau)$ is in fact the defining ideal of X_τ, $\tau \in I_{d,n}$. Thus the ideal generated by Plücker polynomials is in fact the defining ideal of $G_{d,n}$ for the Plücker embedding.*

Remark 5.4.9. Thus X_τ is scheme theoretically (even at the cone level) the intersection of $G_{d,n}$ with all the hyperplanes in $\mathbb{P}\left(\bigwedge^d K^n\right)$ which contain X_τ.

Remark 5.4.10. Because X_τ is an integral scheme for all $\tau \in I_{d,n}$, we have that $I(\tau)$ (and hence I) is a prime ideal for all $\tau \in I_{d,n}$.

Definition 5.4.11. Let $w, \tau, \phi \in I_{d,n}$ such that $w \geq \tau, \phi$, and $p_\tau p_\phi$ nonstandard. Then by the results above, $p_\tau p_\phi$ is equal to a sum of monomials standard on X_w; we call such an equation a *straightening relation*.

Proposition 5.4.12. *Let $w, \tau, \phi \in I_{d,n}$ such that $w \geq \tau, \phi$, and $p_\tau p_\phi$ nonstandard. Let*

$$p_\tau p_\phi = \sum_{(\alpha,\beta)} c_{\alpha,\beta} p_\alpha p_\beta, \quad c_{\alpha,\beta} \in K^*, \alpha \geq \beta \qquad (*)$$

be the straightening relation of $p_\tau p_\phi$ on X_w. Then for each (α, β) appearing in $(*)$, we have $\alpha > \phi, \tau$.

Proof. Among the α's appearing in $(*)$, pick a minimal element and call it α_{min}. We restrict the straightening relation $(*)$ to $X_{\alpha_{min}} \subseteq X_w$. Assume, if possible that $\alpha_{min} \not\geq \tau$, or $\alpha_{min} \not\geq \phi$. In either case, the left hand side of $(*)$ is zero on $X_{\alpha_{min}}$. On the right hand side, at least one term does not go to zero, namely the term containing $p_{\alpha_{min}}$ which is a standard monomial on $X_{\alpha_{min}}$. This contradicts the linear independence of standard monomials (Proposition 5.4.6). Thus our assumption is wrong, and we get $\alpha_{min} \geq \tau, \phi$.

Next we show that $\alpha_{min} > \tau, \phi$. Recall from §5.2.3 that $p_{\underline{i}}$ has weight $-\varepsilon_{i_1} - \ldots - \varepsilon_{i_d}$. Thus by weight considerations, $\tau \dot{\cup} \phi = \alpha \dot{\cup} \beta$ for any pair α, β appearing on the right hand side of $(*)$ (here we are viewing $\tau \dot{\cup} \phi$ and $\alpha \dot{\cup} \beta$ as multisets). Thus if $\alpha_{min} \in \{\tau, \phi\}$, say $\alpha_{min} = \tau$, then the β that appears in the monomial with α_{min} must equal ϕ by weight considerations. This implies $p_\tau p_\phi$ is standard, a contradiction. Therefore $\alpha_{min} > \tau, \phi$. Since this is true for any minimal α, it is true for all α appearing in $(*)$. □

5.5 Unions of Schubert Varieties

Definition 5.5.1. Given $X_{\underline{j}} \subset X_{\underline{i}}$, if $\dim X_{\underline{j}} = \dim X_{\underline{i}} - 1$, then $X_{\underline{j}}$ is called a *Schubert divisor* of $X_{\underline{i}}$.

Lemma 5.5.2. *Let $\tau < w$. There exists a Schubert divisor X_ϕ of X_w such that $X_\tau \subseteq X_\phi$.*

Proof. Let $\tau = (j_1, \ldots, j_d)$ and $w = (i_1, \ldots, i_d)$ be such that $j_1 \leq i_1, \ldots, j_d \leq i_d$. Let l be the smallest integer such that $1 \leq l \leq d$ and $j_l < i_l$. Set $\phi = (i_1, \ldots, i_{l-1}, i_l - 1, i_{l+1}, \ldots, i_d)$. Once we show that $\phi \in I_{d,n}$, the result will follow, clearly. To prove this, we first note that if $l = 1$, then clearly $\phi \in I_{d,n}$. Let then $l > 1$. We shall show that $i_{l-1} < i_l - 1$ (from which it will follow that $\phi \in I_{d,n}$). Assume, if possible that $i_{l-1} = i_l - 1$ (note that $i_{l-1} \leq i_l - 1$); then $j_{l-1} = i_{l-1} = i_l - 1$. This implies $j_l > i_l - 1$, and thus $j_l \geq i_l$, a contradiction (note that $j_l < i_l$). Thus our assumption is wrong, and we get $\phi \in I_{d,n}$. □

Let X_w be a Schubert variety in $G_{d,n}$. Let $\{X_{w_i}\}$ be the set of Schubert divisors in X_w. By Bruhat decomposition, we have

$$\partial X_w = \bigcup X_{w_i},$$

(∂X_w being the boundary of X_w, namely, $\partial X_w = X_w \setminus B \cdot e_w$). In the sequel, we shall denote ∂X_w by H_w. Clearly $X_w \cap \{p_w = 0\} = H_w$, set theoretically. We shall see (cf. Theorem 5.5.8) that this equality is in fact scheme-theoretic; i.e., $I(H_w) = \langle p_w \rangle$ ($\subset S/I(w)$); here $I(H_w)$ is the ideal in $S/I(w)$ so that the corresponding quotient is $K[H_w]$.

We first describe a standard monomial basis for a union of Schubert varieties.

Definition 5.5.3. Let $Y = \bigcup_i X_{\tau_i}$, where $\tau_i \in I_{d,n}$. A monomial $p_{\theta_1} \ldots p_{\theta_m}$ is *standard on Y* if it is standard on X_{τ_i} for some i.

Theorem 5.5.4. *Standard monomials on $Y = \bigcup_{i=1}^r X_{\tau_i}$ give a basis for $R(Y)$ (here $R(Y) = K[Y]$).*

Proof. We first show linear independence. If possible, let

$$\sum_{i=1}^{s} a_i F_i = 0, \quad F_i = p_{\tau_{i1}} \ldots p_{\tau_{ir_i}}, a_i \neq 0 \qquad (*)$$

be a nontrivial relation among standard monomials on Y. For simplicity, denote τ_{11} by τ. Then, restricting $(*)$ to X_τ, we obtain a nontrivial relation among standard monomials on X_τ, which is a contradiction (note that for any i, $F_i|_{X_\tau}$ is either 0 or remains standard on X_τ; further, $F_1|_{X_\tau}$ is nonzero). This proves the linear independence of standard monomials on Y.

The generation of $R(Y)$ by standard monomials follows from the generation of $K[G_{d,n}]$ by standard monomials and the canonical surjection $K[G_{d,n}] \to R(Y)$. □

Remark 5.5.5. The union of two reduced schemes is again reduced. On the other hand, the intersection of two reduced schemes need not be reduced. If $W = X \cap Y$, then $I_W = I_X + I_Y$. The fact that I_X and I_Y are radical ideals need not necessarily imply that I_W is a radical ideal.

However, as a consequence of standard monomial theory, we shall now show that an intersection of Schubert varieties is reduced.

Remark 5.5.6. It is clear that set-theoretically, the intersection of two Schubert varieties is a union of Schubert varieties. In fact, it is seen easily that for $\underline{i}, \underline{j} \in I_{d,n}$, we have $X_{\underline{i}} \cap X_{\underline{j}} = X_\tau$, where $\tau = (a_1, \ldots, a_d), a_k = \min\{i_k, j_k\}$, $1 \leq k \leq d$.

Theorem 5.5.7. *Let X_1, X_2 be two Schubert varieties in $G_{d,n}$. Then $X_1 \cap X_2$ is reduced.*

Proof. Let $X_j = X_{w_j}$, $R(w_j) = R/I_j$, R being $K[G_{d,n}]$. Let $A = R/I$, where $I = I_1 + I_2$. We need to show that $\sqrt{I} = I$.

Now I consists of linear sums $\sum a_j F_j$, F_j standard; where each F_j is zero either on X_1 or X_2; i.e., each F_j starts with $p_{\theta_{j_1}}$ where

$$\text{either } \theta_{j_1} \not\leq w_1 \text{ or } \theta_{j_1} \not\leq w_2. \qquad (*)$$

On the other hand, if $A' = R/\sqrt{I}$, then $A' = K[X_\tau]$, where $X_1 \cap X_2 = X_\tau$.

Therefore, using the standard monomial basis for X_τ, we obtain that $\sqrt{I}$ (the kernel of $\pi : R \to A'$) consists of all $f = \sum c_k f_k$ where f_k starts with

$$p_{\delta_{k1}}, \text{ where } \delta_{k1} \not\leq \tau. \qquad (**)$$

Equivalently, either $\delta_{k1} \not\leq w_1$ or $\delta_{k1} \not\leq w_2$. Thus, comparing $(*)$ and $(**)$, we obtain that $I = \sqrt{I}$. □

Theorem 5.5.8 (Pieri's formula). *Let $\tau \in I_{d,n}$, and let $X_{\tau_1}, \ldots, X_{\tau_r}$ be the Schubert divisors of X_τ. Then*

$$X_\tau \cap \{p_\tau = 0\} = \bigcup_{i=1}^{r} X_{\tau_i},$$

scheme theoretically.

Proof. Let $H_\tau = \bigcup_{i=1}^r X_{\tau_i}(\subset X_\tau)$, and let I_{H_τ} be the defining ideal of H_τ in $R(\tau)$. Clearly $p_\tau|_{H_\tau} = 0$ since $\tau \not\leq \tau_i$, $\forall\, 1 \leq i \leq r$. Therefore $(p_\tau) \subseteq I_{H_\tau}$.

We now show $I_{H_\tau} \subseteq (p_\tau)$. Let $f \in I_{H_\tau}$. Write f as a sum of standard monomials on X_τ. Let $f = \sum a_i F_i + \sum b_j G_j$, where each F_i is a standard monomial in $R(\tau)$ starting with p_τ, and each G_j is a standard monomial $R(\tau)$ starting with $p_{\theta_j}, \theta_j < \tau$. Since $p_\tau \in I_{H_\tau}$, we have $\sum a_i F_i \in I_{H_\tau}$, and thus $\sum b_j G_j \in I_{H_\tau}$. Thus $\sum b_j G_j = 0$ on H_τ. But $\sum b_j G_j$ being a sum of standard monomials on H_τ, we conclude by Theorem 5.5.4 that each $b_j = 0$. Therefore $f \in (p_\tau)$, and the result follows. □

5.5.1 *The Picard Group*

Recall that Pic X, the Picard group of an algebraic variety is the group of isomorphism classes of line bundles on X.

Theorem 5.5.9 (cf. Theorem 6.1 of [77]). *The Picard group of the Grassmannian is isomorphic to the group of integers, i.e.,*

$$Pic\, G_{d,n} \cong \mathbb{Z}.$$

More generally, the Picard group of a (nontrivial) Schubert variety is $\mathbb{Z}$.

In fact, we have that any line bundle on $G_{d,n}$ (as well as a Schubert variety) is of the form $L^{\otimes m}$, for some integer m, where L is the pull-back of $\mathcal{O}_{\mathbb{P}^N}(1)$ via the Plücker embedding (here, $N = \binom{n}{d} - 1$).

5.6 Vanishing Theorems

In this section, we prove the vanishing of $H^i(X_\tau, L^m)$, $i \geq 1, m \geq 0$, for $\tau \in I_{d,n}$. Further, we show that the standard monomials on X_τ give a basis for $H^0(X_\tau, L^m)$, $m \geq 0$.

Note that for a projective variety $X \hookrightarrow \mathbb{P}^n$, denoting $K[X]$, the homogeneous coordinate ring of X by R, and $\mathcal{O}_{\mathbb{P}^n}(1)$ by L, we have for $m \geq 0$, the image of the restriction map $\phi_m : H^0(\mathbb{P}^n, L^m) \to H^0(X, L^m)$ is precisely R_m (the m-th graded piece of R). Thus as a consequence of the main results of this section, we shall see (cf. Corollary 5.6.5) that the inclusion $R(\tau)_m \subseteq H^0(X_\tau, L^m)$ is in fact an equality, for $m \geq 0$; thus

$$K[X_\tau] = \bigoplus_{m \in \mathbb{Z}^+} H^0(X_\tau, L^m),\ \tau \in I_{d,n}.$$

We first prove some preparatory Lemmas to be used in proving the main results.

Let X be a union of Schubert varieties in $G_{d,n}$. Let $S(X,m)$ be the set of all standard monomials on X of degree m, and $s(X,m)$ be the cardinality of $S(X,m)$. The proof of the following lemma follows easily from the results of the previous sections, and so we leave it to the reader.

Lemma 5.6.1.

1. *Let* $X = X_1 \cup X_2$, *where both* X_1, X_2 *are unions of Schubert varieties. Then*

$$s(X,m) = s(X_1,m) + s(X_2,m) - s(X_1 \cap X_2, m).$$

2. *For* $\tau \in I_{d,n}$, *let* H_τ *be the union of all Schubert divisors as in Theorem 5.5.8. Then*

$$s(X_\tau, m) = s(X_\tau, m-1) + s(H_\tau, m).$$

Let $L = \mathcal{O}_{G_{d,n}}(1)$ (as defined in §2.7); note that L is just the pull-back of $\mathcal{O}_{\mathbb{P}^N}(1)$ via the Plücker embedding. We shall denote the restriction of L to a subvariety of $G_{d,n}$ by just L. For $m \in \mathbb{Z}$, we will denote $\mathcal{O}_{G_{d,n}}(m)$ by L^m. Before moving on, the reader may want to review Definition 3.4.1 concerning $H^i(X,L)$.

Proposition 5.6.2. *Let* r *be an integer such that* $r \le d(n-d)$ $(= \dim G_{d,n})$. *Suppose that all Schubert varieties* X *in* $G_{d,n}$ *of dimension less than or equal to* r *have the following properties:*

1. $H^i(X,L^m) = 0$, $i \ge 1$, $m \ge 0$.
2. $S(X,m)$ *is a basis for* $H^0(X,L^m)$.

Then any union or intersection of Schubert varieties of dimension at most r *has the same properties.*

Proof. We prove the result by induction on r. Suppose the result is true for Schubert varieties of dimension at most $r-1$. Let S_r denote the set of Schubert varieties in $G_{d,n}$ of dimension at most r. Let $Y = \bigcup_{j=1}^{t} X_j$, $X_j \in S_r$. We will also use induction on t. By hypothesis, the result is clearly true for $t = 1$. Let then $t \ge 2$ and suppose the result holds for unions of $t-1$ Schubert varieties. Let $Y_1 = \bigcup_{j=1}^{t-1} X_j$, and $Y_2 = X_t$. We have the short exact sequence

$$0 \to \mathcal{O}_Y \to \mathcal{O}_{Y_1} \oplus \mathcal{O}_{Y_2} \to \mathcal{O}_{Y_1 \cap Y_2} \to 0. \tag{$*$}$$

The first map above sends $f \mapsto (f|_{Y_1}, f|_{Y_2})$; the second map sends $(f,g) \mapsto (f-g)|_{Y_1 \cap Y_2}$. Tensoring $(*)$ with L^m, we have the long exact cohomology sequence (cf. Remark 3.2.13 (4)):

$$\to H^{i-1}(Y_1 \cap Y_2, L^m) \to H^i(Y, L^m) \to H^i(Y_1, L^m) \oplus H^i(Y_2, L^m) \to$$

Now $Y_1 \cap Y_2$ is reduced (cf. Theorem 5.5.7), thus $Y_1 \cap Y_2$ is a union of Schubert varieties of dimension at most $r-1$. Let $i \ge 2$ and $m \ge 0$; by hypothesis, $H^i(Y_2, L^m) = 0$, and by induction hypothesis, $H^i(Y_1, L^m) = 0$ and $H^{i-1}(Y_1 \cap Y_2, L^m) = 0$. Thus by the long exact sequence, $H^i(Y, L^m) = 0$ for $i \geq 2$.

Since Y_2 is a Schubert variety, the hypothesis implies $S(Y_2, m)$ is a basis for $H^0(Y_2, L^m)$. By the induction hypothesis, $H^0(Y_1 \cap Y_2, L^m)$ has standard monomials of degree m as a basis. Therefore

$$H^0(Y_1, L^m) \oplus H^0(Y_2, L^m) \to H^0(Y_1 \cap Y_2, L^m)$$

is a surjection (note that any standard monomial of degree m on $Y_1 \cap Y_2$ may be realized as the restriction to $Y_1 \cap Y_2$ of a standard monomial of degree m on Y_2). This implies (in view of the long exact sequence) that the map $H^0(Y_1 \cap Y_2, L^m) \to H^1(Y, L^m)$ is the zero map. Also, the map $H^1(Y, L^m) \to H^1(Y_1, L^m) \oplus H^1(Y_2, L^m)$ is the zero map (since, by induction, $H^1(Y_1, L^m) = 0$ and by hypothesis, $H^1(Y_2, L^m) = 0$). Hence, we obtain (in view of the long exact sequence) that $H^1(Y, L^m) = 0$, and the proof of (1) is complete.

To see (2), we begin with the short exact sequence

$$0 \to H^0(Y, L^m) \to H^0(Y_1, L^m) \oplus H^0(Y_2, L^m) \to H^0(Y_1 \cap Y_2, L^m) \to 0.$$

Thus, $\dim H^0(Y, L^m) =$

$$\dim H^0(Y_1, L^m) + \dim H^0(Y_2, L^m) - \dim H^0(Y_1 \cap Y_2, L^m).$$

By the induction hypothesis and Lemma 5.6.1, $\dim H^0(Y, L^m) = s(Y, m)$. By linear independence of standard monomials, the result follows (as remarked above, note that $R(\tau)_m \subseteq H^0(X_\tau, L^m)$, and thus $S(X_\tau, m) \subseteq H^0(X_\tau, L^m)$). □

Proposition 5.6.3. *Let X be a Schubert variety in $G_{d,n}$. Let $Y = \bigcup_{j=1}^t X_j$ be a union of Schubert divisors in X. Suppose $H^i(X_j, L^m) = 0$, for $m < 0$, $0 \leq i \leq \dim Y - 1$ $(= \dim X_j - 1)$. Then $H^i(Y, L^m) = 0$, for $m < 0$, $0 \leq i \leq \dim Y - 1$.*

Proof. We prove the result by induction on t and $\dim Y (= \dim X - 1)$. As in the previous proof, set $Y_1 = \bigcup_{j=1}^{t-1} X_j$, $Y_2 = X_t$, and consider the long exact cohomology sequence

$$\to H^{i-1}(Y_1 \cap Y_2, L^m) \to H^i(Y, L^m) \to H^i(Y_1, L^m) \oplus H^i(Y_2, L^m) \to$$

By Remark 5.5.6 and the definition of a Schubert divisor, we have that $X_j \cap X_t$ is irreducible of codimension 1 in X_t, for $1 \leq j \leq t-1$; and $Y_1 \cap Y_2 = \bigcup_{j=1}^{t-1} X_j \cap X_t$. Hence by induction on $\dim Y$, we have $H^i(Y_1 \cap Y_2, L^m) = 0$ for $m < 0$, $0 \leq i \leq \dim Y_1 - 2 = \dim Y - 2$. Further, by induction on t, $H^i(Y_1, L^m) = 0$, $m < 0$, $0 \leq i < \dim Y_1$; and by hypothesis, $H^i(Y_2, L^m) = 0$, $m < 0$, $0 \leq i < \dim Y_1$. The required result now follows. □

Theorem 5.6.4 (Vanishing Theorems). *Let X be a Schubert variety in $G_{d,n}$.*

1. *$H^i(X, L^m) = 0$ for $i \geq 1$, $m \geq 0$.*
2. *$H^i(X, L^m) = 0$ for $0 \leq i < \dim X$, $m < 0$.*
3. *The set $S(X, m)$ is a basis for $H^0(X, L^m)$.*

Proof. We first prove results (1) and (3) (by induction on m and $\dim X$). If $\dim X = 0$, then X is a point, and the result clearly follows. Assume now that $\dim X \geq 1$. Let $X = X_\tau$, and let $X_1, \ldots, X_s$ be all the Schubert divisors of X. Define $Y = \bigcup_{i=1}^{s} X_i$. By Pieri's formula 5.5.8, $Y = X_\tau \cap \{p_\tau = 0\}$. Thus we have the short exact sequence

$$0 \to \mathcal{O}_X(-1) \to \mathcal{O}_X \to \mathcal{O}_Y \to 0$$

where the map $\mathcal{O}_X(-1) \to \mathcal{O}_X$ is given by multiplication by p_τ (note that $\mathcal{O}_X(-1) = L^{-1}$). Tensoring with L^m, we obtain the following long exact cohomology sequence

$$\to H^{i-1}(Y, L^m) \to H^i(X, L^{m-1}) \to H^i(X, L^m) \to H^i(Y, L^m) \to \qquad (*)$$

Let $m \geq 0$. Then by the induction hypothesis on $\dim X$, we may invoke Proposition 5.6.2, which implies $H^i(Y, L^m) = 0$ for $i \geq 1$. Thus the sequence $0 \to H^i(X, L^{m-1}) \to H^i(X, L^m)$ is exact for $i \geq 2$. If $i = 1$, again the induction hypothesis implies the surjectivity of $H^0(X, L^m) \to H^0(Y, L^m)$. This implies that the map $H^0(Y, L^m) \to H^1(X, L^{m-1})$ is the zero map, and hence the sequence $0 \to H^1(X, L^{m-1}) \to H^1(X, L^m)$ is exact. Thus, for $i \geq 1$ and $m \geq 0$, we have the exact sequence

$$0 \to H^i(X, L^{m-1}) \to H^i(X, L^m).$$

However, by [81, §63], there exists a m_0 such that for every $m \geq m_0$, $H^i(X, L^m) = 0$ for all $i \geq 1$. Therefore

$$H^i(X, L^m) = 0 \text{ for } i \geq 1,\ m \geq 0,$$

completing part (1).

For $i = 0$, we have the short exact sequence

$$0 \to H^0(X, L^{m-1}) \to H^0(X, L^m) \to H^0(Y, L^m) \to 0$$

which implies

$$\dim H^0(X, L^m) = \dim H^0(X, L^{m-1}) + \dim H^0(Y, L^m).$$

For $m \geq 1$, the induction hypothesis on m implies that $\dim H^0(X, L^{m-1}) = s(X, m-1)$. (We note that for $m = 1$, $H^0(X, L^{m-1}) = H^0(X, \mathcal{O}_X) = K$, and thus $\dim H^0(X, L^{m-1}) = s(X, m-1)$.) On the other hand, the induction on $\dim X$, combined with Proposition 5.6.2, implies that $\dim H^0(Y, L^m) = s(Y, m)$. Therefore

$$\dim H^0(X, L^m) = s(X, m-1) + s(Y, m).$$

This together with Lemma 5.6.1 implies, $\dim H^0(X, L^m) = s(X, m)$, and result (3) follows by linear independence of standard monomials on X (note that $R(X)_m \subseteq H^0(X, L^m)$, and thus $S(X, m) \subseteq H^0(X, L^m)$).

Finally, it remains to prove part (2). Once again we use the long exact sequence in (*). Proceeding by induction on $\dim X$, by Propositions 5.6.2 and 5.6.3, we have $H^i(Y, L^m) = 0$ for $m \in \mathbb{Z}$, $1 \leq i \leq \dim Y - 1$. Thus we obtain the exact sequence

$$0 \to H^i(X, L^{m-1}) \to H^i(X, L^m),\ 2 \leq i < \dim X.$$

But again by [81, §63], $H^i(X, L^m) = 0$ for m sufficiently large, and thus $H^i(X, L^m) = 0$ for $2 \leq i < \dim X$, $m \in \mathbb{Z}$.

It remains to prove (2) for $i = 0, 1$ and $m < 0$. Let $m < 0$, then $H^0(Y, L^m) = 0$ by the induction hypothesis and Proposition 5.6.3. Thus we have the exact sequence $0 \to H^1(X, L^{m-1}) \to H^1(X, L^m)$, $m < 0$. Thus we obtain the following chain of inclusions

$$H^1(X, L^{m-1}) \subseteq H^1(X, L^m) \subseteq \ldots \subseteq H^1(X, L^{-1}).$$

But now the isomorphism $H^0(X, \mathcal{O}_X) \cong H^0(Y, \mathcal{O}_Y)$ together with the fact that $H^1(X, \mathcal{O}_X) = 0$ implies that $H^1(X, L^{-1}) = 0$. By the inclusions above, we conclude

$$H^1(X, L^m) = 0,\ m < 0.$$

Next, the fact that $H^0(Y, L^m) = 0$ for $m < 0$, implies

$$(\dagger) \qquad H^0(X, L^{m-1}) \cong H^0(X, L^m), m < 0$$

Again, the isomorphism $H^0(X, \mathcal{O}_X) \cong H^0(Y, \mathcal{O}_Y)$ implies that $H^0(X, L^{-1}) = 0$. This together with (†) implies that $H^0(X, L^m) = 0$ for $m < 0$, thus completing the proof. □

Corollary 5.6.5.

1. $K[G_{d,n}] = \bigoplus_{m \in \mathbb{Z}^+} H^0(G_{d,n}, L^m)$.
2. $K[X_\tau] = \bigoplus_{m \in \mathbb{Z}^+} H^0(X_\tau, L^m)$.

Chapter 6
Further Geometric Properties of Schubert Varieties

In this chapter, using standard monomial basis, we prove Cohen–Macaulayness and normality for Schubert varieties; these geometric properties are proved even for the cones over Schubert varieties. We also give a characterization of arithmetically factorial Schubert varieties. The chapter also includes determination of the singular locus of a Schubert variety, and the tangent space at a singular point of a Schubert variety.

6.1 Cohen–Macaulay

In this section, we give two proofs that all Schubert varieties are arithmetically Cohen–Macaulay; i.e., the cone over a Schubert variety is Cohen–Macaulay. We begin with the following lemma concerning a general projective variety.

Lemma 6.1.1. *Let $X \subset \mathbb{P}^n$ be a closed subscheme. Then $\hat{X}$ is Cohen–Macaulay if and only if $\hat{X}$ is Cohen–Macaulay at its vertex.*

Proof. The first direction is clear; we prove the reverse implication. Suppose the vertex (0) is a Cohen–Macaulay point of $\hat{X}$. Then there exists a neighborhood U of (0) in $\hat{X}$ that is Cohen–Macaulay. Clearly, each fiber of the projection $\pi : \hat{X} \setminus (0) \to X$ intersects U. But each fiber is also a K^* orbit. Thus each point of $\hat{X}$ (other than (0)) is in an orbit of a Cohen–Macaulay point of $\hat{X}$. Therefore, each point is Cohen–Macaulay. □

As a result of the lemma above, whenever we show that a variety is arithmetically Cohen–Macaulay, we will simply show that the cone over the variety is Cohen–Macaulay at its vertex.

Next we prove a lemma concerning regular sequences.

V. Lakshmibai, J. Brown, *The Grassmannian Variety*,
Developments in Mathematics 42, DOI 10.1007/978-1-4939-3082-1_6

Lemma 6.1.2. *Let A be a Noetherian local domain with maximal ideal $\mathfrak{m}$. Let I and J be ideals in A such that depth $A/(I+J) \geq d-1$, depth $A/I \geq d$, and depth $A/J \geq d$. Suppose $f_1, \ldots, f_d \in \mathfrak{m}$ are such that $f_1, \ldots, f_d (mod\, I)$ (resp. $(mod\, J)$) form an A/I (resp. A/J) regular sequence, and $f_1, \ldots, f_{d-1} (mod\, I+J)$ is an $A/(I+J)$-regular sequence. Then $f_1, \ldots, f_d (mod\, I \cap J)$ is an $A/I \cap J$-regular sequence.*

Proof. Since the conclusion of the lemma is only about $A/I \cap J$, we may assume that $I \cap J = (0)$. We will prove the result by induction on d, the length of the sequence $\{f_1, \ldots, f_d\}$. Note that for $d = 1$, the result holds, since f_1 is a nonzero divisor in A. Now we suppose the result holds for sequences of length $d-1$.

The assumption that $I \cap J = (0)$ implies an inclusion

$$\theta : A \hookrightarrow A/I \times A/J, a \mapsto (a\,(mod\, I), a\,(mod\, J)).$$

We first observe the following:

$$(\dagger) \quad \theta(A) = \{(a_1\,(mod\, I), a_2\,(mod\, J)) \in A/I \times A/J \mid a_1 \equiv a_2\,(\text{mod}\, I+J)\}.$$

(If $(a_1\,(mod\, I), a_2\,(mod\, J)) \in A/I \times A/J$ is such that $a_1 \equiv a_2\,(\text{mod}\, I+J)$, say, $a_1 - a_2 = x + y$, for some $x \in I, y \in J$, then letting $z = a_1 - x = a_2 + y$, we have, $\theta(z) = (a_1\,(mod\, I), a_2\,(mod\, J))$.)

Let H denote the ideal in $A/I \times A/J$ generated by (f_1, f_1). For a subset S of A, we shall denote by $\langle S \rangle$, the ideal generated by S. Towards proving the result, we first claim that

$$(*) \qquad \langle f_1 \rangle = \langle f_1, I \rangle \cap \langle f_1, J \rangle .$$

We prove the claim $(*)$ in two steps: first we show that $\theta(\langle f_1, I\rangle \cap \langle f_1, J\rangle) = \theta(A) \cap H$; next, we show that $\theta(\langle f_1 \rangle) = \theta(A) \cap H$. Once, these two steps are established, the claim $(*)$ will follow in view of the injectivity of θ.

(I) Towards establishing the first step , we observe that

$$(a_1\,(mod\, I), a_2\,(mod\, J)) \in \theta(A) \cap H$$

if and only if (a) and (b) below hold:

(a) $a_1 = cf_1 + x_0, c \in A, x_0 \in I, a_2 = df_1 + y_0, d \in A, y_0 \in J$.
(b) $a_1 - a_2 = x + y$, for some $x \in I, y \in J$ (cf. $(\dagger)$ above).

Clearly, (a) and (b) hold if and only if $z (= a_1 - x = a_2 + y) \in \langle f_1, I \rangle \cap \langle f_1, J \rangle$ and $\theta(z) = (a_1\,(mod\, I), a_2\,(mod\, J))$. Hence we obtain that $\theta(\langle f_1, I\rangle \cap \langle f_1, J\rangle) = \theta(A) \cap H$.

(II) Next, we shall show that $\theta(\langle f_1 \rangle) = \theta(A) \cap H$. Clearly, $\theta(\langle f_1 \rangle) \subseteq \theta(A) \cap H$. Now we show the reverse inclusion.

Let $(a_1\,(mod\,I), a_2\,(mod\,J)) \in \theta(A) \cap H$. Then $a_1 = f_1\alpha + x, a_2 = f_1\beta + y$, for some $\alpha, \beta \in A, x \in I, y \in J$. Since $(a_1\,(mod\,I), a_2\,(mod\,J)) \in \theta(A)$, we have $a_1 - a_2 \in I + J$ (cf. (†) above), and hence $f_1\alpha - f_1\beta \in I + J$. By hypothesis, f_1 is not a zero divisor in $A/(I + J)$, hence $\alpha - \beta \in I + J$, say, $(\alpha\,(mod\,I), \beta\,(mod\,J)) = \theta(\gamma)$, for a unique $\gamma \in A$. This implies that $(a_1\,(mod\,I), a_2\,(mod\,J)) = (f_1\alpha\,(mod\,I), f_1\beta\,(mod\,J)) = \theta(f_1\gamma)$, and thus $(a_1\,(mod\,I), a_2\,(mod\,J)) \in \theta(\langle f_1\rangle)$. Thus we get $\theta(\langle f_1\rangle) = \theta(A) \cap H$.

(I) and (II) imply $\langle f_1\rangle = \langle f_1, J\rangle \cap \langle f_1, I\rangle$. By hypothesis, $f_2, \ldots, f_d$ (mod $\langle f_1, I\rangle$) (resp. (mod $\langle f_1, J\rangle$)) form an $A/\langle f_1, I\rangle$ (resp. $A/\langle f_1, J\rangle$) -regular sequence, and $f_2, \ldots, f_{d-1}$ (mod $\langle f_1, I + J\rangle$) form an $A/\langle f_1, I + J\rangle$-regular sequence. Hence by induction on d, we obtain that

$$f_2, \ldots, f_d\,(\text{mod } \langle f_1, I\rangle \cap \langle f_1, J\rangle)$$

form an $A/(\langle f_1, I\rangle \cap \langle f_1, J\rangle)$-regular sequence, i.e., $f_2, \ldots, f_d$ (mod $\langle f_1\rangle$) form an $A/\langle f_1\rangle$-regular sequence (cf. (∗) above); this together with the fact that f_1 is a nonzero divisor in A implies the required result. □

Theorem 6.1.3. *Let $\tau \in I_{d,n}$, then X_τ is arithmetically Cohen–Macaulay, i.e., the cone $\widehat{X_\tau}$ is Cohen–Macaulay.*

Proof. Let $\dim X_\tau = t$. For $0 \le j \le t$, let $\{X_{\phi_{kj}}\}_k$ be the set of all Schubert subvarieties of X_τ of codimension j. Set

$$f_j(\tau) = \sum_k p_{\phi_{kj}}.$$

By inducting on t, we shall show that $\{f_0(\tau), \ldots, f_t(\tau)\}$ forms a regular system of parameters at the vertex. If $t = 0$ or $t = 1$, the result is clear. (Note that if $t = 1$, then $\tau = (1, \ldots, d-1, d+1)$, and from the discussion following Corollary 5.3.5, $X_\tau = \mathbb{P}^1$.)

Now let $t > 1$. Let $X_1, \ldots, X_r$ be all the Schubert divisors of X_τ. For $1 \le s \le r$, let $Y_s = \bigcup_{i=1}^{s} X_s$. We have $\dim Y_s = t - 1$. For $0 \le j \le t-1$, set $f_{js} = \sum_k p_{\phi_{kj}}$, where $\{X_{\phi_{kj}}\}$ are the codimension j Schubert subvarieties of Y_s.

Now by inducting on s, we will show that $\{f_{0s}, \ldots, f_{t-1s}\}$ forms a system of parameters at the vertex for $\widehat{Y_s}$. If $s = 1$, then Y_s is a Schubert variety of dimension $t-1$, and the result follows by the inductive hypothesis on t. Let $1 < s \le r$; set $X = X_s$ and $Y = \bigcup_{i=1}^{s-1} X_i$, so that $Y_s = X \cup Y$. Note that $X \cap Y$ is a union of Schubert subvarieties of X_τ of codimension 2 (in X_τ). (If $\tau = (a_1, \ldots, a_d)$, then the intersection of two distinct Schubert divisors will have index $(a_1, \ldots, a_i - 1, \ldots a_j - 1, \ldots a_d)$.) Therefore, $\widehat{Y_s}$ is Cohen–Macaulay at its vertex by induction on s, in view of Lemma 6.1.2.

Now we recall that by Pieri's formula (5.5.8),

$$Y_r = \bigcup_{i=1}^{r} X_i = X_\tau \cap \{p_\tau = 0\}.$$

From the discussion above, we have that $\{f_1(X_\tau),\ldots,f_t(\tau)\}$ is a system of parameters at the vertex for $\widehat{Y}_r$. The stalk $\mathcal{O}_{\widehat{Y}_r,0}$ is $K[X_\tau]_{(0)}/\langle p_\tau\rangle$. Note that $f_0(\tau) = p_\tau$ (and p_τ is a nonzero divisor in $K[X_\tau]_{(0)}$), and thus $\{f_0(\tau),\ldots,f_t(X_\tau)\}$ is a system of parameters at the vertex of $\widehat{X_\tau}$. Therefore X_τ is arithmetically Cohen–Macaulay. □

We will now provide a second proof that Schubert varieties are arithmetically Cohen–Macaulay. This proof hinges on the following lemma which makes use of local cohomology (cf. Definition 3.4.2).

Lemma 6.1.4. *Let $A = \bigoplus_{n\geq 0} A_n$ be a Noetherian graded ring with $A_0 = K$. Let I, J be two homogeneous ideals of A such that $\dim A/I = \dim A/J = \dim(A/(I+J)) + 1$ (let $\dim A/I = d$). If A/I, A/J, and $A/(I+J)$ are Cohen–Macaulay, then $A/I\cap J$ is Cohen–Macaulay of dimension d.*

Proof. Let $\mathfrak{m} = \bigoplus_{n\geq 1} A_n$. Since $\mathfrak{m}$ is the (irrelevant) maximal ideal of A, by Proposition 2.1.17, it suffices to show that $(A/I\cap J)_\mathfrak{m}$ is Cohen–Macaulay. Consider the exact sequence

$$0 \to (A/I\cap J)_\mathfrak{m} \to (A/I)_\mathfrak{m} \oplus (A/J)_\mathfrak{m} \to (A/(I+J))_\mathfrak{m} \to 0,$$

where $a \in (A/I\cap J)_\mathfrak{m}$ maps to $(a,a) \in (A/I)_\mathfrak{m} \oplus (A/J)_\mathfrak{m}$ and $(f,g) \in (A/I)_\mathfrak{m} \oplus (A/J)_\mathfrak{m}$ maps to $(f-g) \in (A/(I+J))_\mathfrak{m}$. Regarding the objects above as A-modules and taking local cohomology, we have the following long exact sequence,

$$\cdots \to H^{i-1}_\mathfrak{m}((A/(I+J))_\mathfrak{m}) \to H^i_\mathfrak{m}((A/I\cap J)_\mathfrak{m}) \to H^i_\mathfrak{m}((A/I)_\mathfrak{m}) \oplus H^i_\mathfrak{m}((A/J)_\mathfrak{m}) \to \cdots.$$

By Corollary 3.4.4, $H^i_\mathfrak{m}((A/I)_\mathfrak{m}) = H^i_\mathfrak{m}((A/J)_\mathfrak{m}) = 0$ for $i \neq d$, and $H^i_\mathfrak{m}((A/(I+J))_\mathfrak{m}) = 0$ for $i \neq d-1$. Therefore $H^i_\mathfrak{m}((A/I\cap J)_\mathfrak{m}) = 0$ if $i < d$, implying

$$d \leq \operatorname{depth}(A/I\cap J)_\mathfrak{m} \leq \dim(A/I\cap J)_\mathfrak{m} \leq d.$$

It follows that $(A/I\cap J)_\mathfrak{m}$ is Cohen–Macaulay. □

Proposition 6.1.5. *Let $\tau \in I_{d,n}$, and let X be a union of some collection of Schubert divisors of X_τ, $X = \bigcup_{i=1}^s X_i$. Suppose that for $w \in I_{d,n}$ with $\dim X_w < \dim X_\tau$, X_w is arithmetically Cohen–Macaulay. Then $\hat{X}$ is Cohen–Macaulay.*

Proof. Let $\dim X_\tau = t$, and let $Y = \bigcup_{i=1}^{s-1} X_i$ and $Z = X_s$ so that $X = Y \cup Z$. By induction on s, we have that $\hat{Y}$ is Cohen–Macaulay. By hypothesis, $\hat{Z}$ is Cohen–Macaulay; and by induction on t, the cone over $Z\cap Y$ is Cohen–Macaulay. Therefore, by Lemma 6.1.4, it follows that $\hat{X}$ is Cohen–Macaulay. □

We now give our second proof that Schubert varieties are arithmetically Cohen–Macaulay.

Theorem 6.1.6. *Let $\tau \in I_{d,n}$, then X_τ is arithmetically Cohen–Macaulay, i.e., the cone $\widehat{X_\tau}$ is Cohen–Macaulay.*

Proof. Let $\dim X_\tau = t$, and we proceed by induction on t. Let H be the union of all Schubert divisors of X_τ. We have $K[H]_{(0)} = K[X_\tau]_{(0)}/\langle p_\tau \rangle$. By the induction hypothesis, and by Proposition 6.1.5, we have that $K[H]_{(0)}$ is Cohen–Macaulay. Since p_τ is a nonzero divisor in $K[X_\tau]_{(0)}$, we have that $K[X_\tau]_{(0)}$ is also Cohen–Macaulay. □

6.2 Lemmas on Normality and Factoriality

In the first part of this section, we discuss a necessary and sufficient condition for a Noetherian ring to be normal. The results of this section will be used in the following section to prove the normality of Schubert varieties.

Throughout this section, R denotes a Noetherian ring.

Definition 6.2.1. We define the following conditions on R. For any integer $i \geq 0$,

1. R satisfies (R_i) if R_P is regular for every prime ideal P in R with $\mathrm{ht}(P) \leq i$.
2. R satisfies (S_i) if depth $R_P \geq \min\{\mathrm{ht}(P), i\}$ for every prime ideal P in R.

Note that if R satisfies (R_i) or (S_i), then R satisfies (R_j) or (S_j) for $0 \leq j \leq i$. We also note that R satisfying (S_i) for all $i \geq 0$ is equivalent to R being Cohen–Macaulay.

Serre's criterion on normality, which we shall prove in this section, is that R being normal is equivalent to R satisfying (R_1) and (S_2). In this section, we also present a proof of the fact that R being reduced is equivalent to R satisfying (R_0) and (S_1).

Definition 6.2.2. Let I be an ideal in R, and P a prime ideal in R, then P is a *prime divisor* of I if there exists $x \in R \setminus I$ such that $P = \{p \in R \mid px \in I\}$.

Clearly, if P is a prime divisor of I, then $P \supset I$.

Remark 6.2.3. A prime divisor of I is also called an *associated prime of* R/I, and the set of prime divisors of I is denoted $Ass(R/I)$. For $I = (0)$, the zero ideal, a prime divisor of I is called an *associated prime of* R, and the set of prime divisors of (0) is denoted $Ass(R)$.

Definition 6.2.4. Given an ideal I in R, the set of all prime ideals P in R such that the localization $(R/I)_P$ is nonzero is called *the support of* R/I, and is denoted $\mathrm{Supp}(R/I)$.

Theorem 6.2.5 (cf. [72], Theorem 6.5). *Let R be a Noetherian ring, and I an ideal in R. We have*

- *$Ass(R/I)$ is a finite set.*
- *$Ass(R/I) \subset Supp(R/I)$*
- *The set of minimal elements of $Ass(R/I)$ and $Supp(R/I)$ coincide.*
- *$Supp(R/I)$ is the set of prime ideals containing I.*

As an immediate consequence of the above theorem, we have the following.

Corollary 6.2.6. *For an ideal I, the minimal prime divisors of I are precisely the minimal prime ideals in R that contain I.*

To facilitate the following discussion, we list some facts concerning $\mathrm{Ass}(R)$. For proofs, the reader may refer to [72].

Fact 1. The set of zero divisors in R is equal to $\bigcup_{P\in Ass(R)} P$.
Fact 2. Let $(0) = \bigcap_{1\le i\le r} Q_i$ be an irredundant (i.e., for any i, $1 \le i \le r$, $(0) \neq \bigcap_{j\neq i} Q_j$) primary decomposition of (0). Further, let Q_i be P_i-primary (namely, $\sqrt{(Q_i)} = P_i$, and P_i is a prime ideal). Then $\mathrm{Ass}(R) = \{P_i, 1 \le i \le r\}$
Fact 3. If R is reduced, then $\mathrm{Ass}(R) = \{P_i, 1 \le i \le s\}$, the set of all minimal prime ideals in R.
Fact 4. Let P be a prime ideal in R. We have that $P \in Ass(R)$ if and only if $PR_P \in Ass(R_P)$ if and only if depth $R_P = 0$.
Fact 5. The following are equivalent for R:

- (S_1) holds.
- For every prime ideal P with height $P \ge 1$, we have depth $R_P \ge 1$.
- Associated prime ideals have height 0.
- Associated prime ideals are minimal.

Lemma 6.2.7. *Let R be a Noetherian ring. Then R satisfies (R_0) and (S_1) if and only if R is reduced.*

Proof. $\Rightarrow$: We have, by Fact 5, $\mathrm{Ass}(R) = \{$minimal prime ideals in $R\} = \{P_i, 1 \le i \le s\}$, say. Then in view of (R_0), we obtain that for $1 \le i \le s$, R_{P_i} is a zero-dimensional regular local ring, and hence is a field.

Let $S = R \setminus \bigcup_{P\in Ass(R)}$ (={nonzero divisors in R}). Now the maximal ideals in $S^{-1}R$ are precisely $S^{-1}P_i$, $1 \le i \le s$, and the localizations $(S^{-1}R)_{S^{-1}P_i} (= R_{P_i})$ being fields are reduced, and hence $S^{-1}R$ is reduced. This together with the inclusion $R \hookrightarrow S^{-1}R$ implies that R is reduced.

$\Leftarrow$: We have, by Fact 3, $\mathrm{Ass}(R) = \{$minimal prime ideals in $R\}$, and hence (S_1) holds (in view of Fact 5).

For a prime ideal P of height 0, we have that R_P is a reduced zero-dimensional local ring, and hence is a field. From this (R_0) follows. □

Fact 6. Let R be reduced; further, let S, P_i, $1 \le i \le s$ be as in the proof of Lemma 6.2.7. Let $K(R/P_i)$ be the quotient field of the domain R/P_i. We have

$$S^{-1}R = \prod_{i=1}^{s} K(R/P_i) = \prod_{i=1}^{s} R_{P_i} \qquad (*)$$

This follows by observing

1. $\bigcap_{1\le i\le s} S^{-1}P_i = (0)$,
2. $S^{-1}R/S^{-1}P_i = K(R/P_i)$,

and using the Chinese Remainder Theorem, included below.

> **Chinese Remainder Theorem:** Let Q_i, $1 \leq i \leq s$ be ideals in a (commutative) ring A such that $Q_i + Q_j = A, i \neq j$. Then $A/\bigcap_{i=1}^{s} Q_i = \prod_{i=1}^{s} A/Q_i$.

Note: As seen in the proof of Lemma 6.2.7, $\{S^{-1}P_i,\ 1 \leq i \leq s\}$ is precisely the set of maximal ideals in $S^{-1}R$, and hence with $A = S^{-1}R$, $Q_i = S^{-1}P_i$, $1 \leq i \leq s$, the hypothesis in the Chinese Remainder Theorem is satisfied; also, R_{P_i} being a field (as seen in the proof of the lemma), we have, $R_{P_i} = K(R/P_i)$, $1 \leq i \leq s$.

Lemma 6.2.8. *Let R be an integral domain. Then R satisfies (S_2) if and only if every prime divisor of a nonzero principal ideal in R has height 1.*

Proof. Let R be an integral domain that satisfies (S_2), and let P be a prime divisor of $\langle a\rangle$ for some nonzero $a \in R$. Thus $a \in P$, and there exists $x \in R$ such that $Px \subset \langle a\rangle$. We also note that $x \notin \langle a\rangle$ (otherwise $P = R$ which is not the case). Therefore, $\{a\}$ is a maximal regular sequence in R_P, because any other element of PR_P will be a zero divisor in $R_P/\langle a\rangle$. Thus depth $R_P = 1$. Because R satisfies (S_2), we have $\mathrm{ht}(P) \leq 1$. Since R is an integral domain, (0) is a prime ideal contained in P, therefore $\mathrm{ht}(P) = 1$.

To prove the converse, let R be an integral domain such that every prime divisor of a nonzero principal ideal has height 1. Fix such a prime divisor P in R. Then by hypothesis, $\mathrm{ht}(P) = 1$; on the other hand, by Remark 2.1.13, depth $R_P \geq 1$ (since R is a domain, R_P is also a domain). Thus, for such a prime divisor P, we have depth $R_P \geq \min\{\mathrm{ht}(P), 2\}$.

To complete the proof of the converse, suppose P is a prime ideal in R that is not the prime divisor of a principal ideal. If $P \supset \langle a\rangle$, and $x \notin \langle a\rangle$, we have $Px \not\subset \langle a\rangle$. This implies $\{a, x\}$ is a regular sequence in R_P; therefore depth $R_P \geq 2$, and hence depth $R_P \geq \min\{\mathrm{ht}(P), 2\}$, and R satisfies (S_2). □

Lemma 6.2.9. *A principal prime ideal of a Noetherian integral domain R has height 1.*

Proof. Let $P = \langle m\rangle$, and suppose P' is a prime ideal properly contained in P. Thus $m \notin P'$. Let $p_o \in P'$. Since $\langle p_0\rangle \subset P$, there exists $p_1 \in R$ such that $p_0 = mp_1$. Since P' is prime, we obtain that $p_1 \in P'$; and $\langle p_0\rangle \subseteq \langle p_1\rangle$. We could continue in this fashion, finding p_2, etc., producing the chain of ideals

$$\langle p_0\rangle \subseteq \langle p_1\rangle \subseteq \langle p_2\rangle \subseteq \ldots,$$

all ideals contained in P'. Because the ring is Noetherian, there exists n such that $\langle p_n\rangle = \langle p_{n+1}\rangle$. By our construction, we already have $p_n = mp_{n+1}$; there must also exist t such that $p_{n+1} = tp_n$. Therefore $p_n = mtp_n$. If $p_n \neq 0$, then $mt = 1$ (since R is an integral domain), implying m is a unit (which is not possible, since m generates P); thus $p_n = 0$, which implies $p_0 = 0$, and therefore $P' = (0)$. Thus P has height 1. □

Theorem 6.2.10. *If R is a normal Noetherian ring, then R satisfies (R_1) and (S_2).*

Proof. Let R be normal. We will show R satisfies (S_2) using the equivalent condition of Lemma 6.2.8. Let P be a prime divisor of $\langle a\rangle \subset R$, $P = \{p \in R \mid pb \in \langle a\rangle\}$ for some $b \in R$. We must show that P has height 1. Let $\mathfrak{m}$ be the maximal ideal PR_P of R_P. Let $K(R_P)$ be the field of fractions of R_P. For an ideal $I \subset R_P$ define $I^{-1} = \{\alpha \in K(R_P) \mid \alpha I \subset R_P\}$.

Clearly $a \in \mathfrak{m}$, and thus $a^{-1} \notin R_P$. The element ba^{-1} in $K(R_P)$ is an element of $\mathfrak{m}^{-1}$ since $\mathfrak{m} = PR_P = \{p \in R_P \mid pb \in aR_P\}$.

Claim 1: $ba^{-1} \notin R_P$. Suppose $ba^{-1} \in R_P$, then $ba^{-1} = \frac{r}{s}$, where $s \notin P$. Thus $sb = ra$, and this is an equation of elements in R; but this implies $sb \in \langle a\rangle$ and thus $s \in P$, a contradiction. Therefore $ba^{-1} \notin R_P$, and Claim 1 follows.

Claim 2: $ba^{-1}\mathfrak{m} = R_P$. If possible, let us assume that $ba^{-1}\mathfrak{m} \subset R_P$. Now, $ba^{-1}\mathfrak{m}$ being an ideal in R_P, we have that $ba^{-1}\mathfrak{m} \subseteq \mathfrak{m}$. Let $\mathfrak{m}$ be generated by $m_1, \ldots, m_n$; i.e., $\mathfrak{m} = m_1R_P + \cdots + m_nR_P$. Thus, for any $1 \le i \le n$, we have $ba^{-1}m_i = \sum_{j=1}^n r_{ij}m_j, r_{ij} \in R_P$, yielding the equation

$$\sum_{j=1}^{n} \left(ba^{-1}\delta_{ij} - r_{ij}\right) m_j = 0$$

where $\delta_{ij} = 0$ unless $i = j$ in which case $\delta_{jj} = 1$. Writing this system of equations in matrix form, we have

$$A \begin{bmatrix} m_1 \\ \vdots \\ m_n \end{bmatrix} = 0$$

where the $(i,j)^{th}$ entry of A is $(ba^{-1}\delta_{ij} - r_{ij})$. If we multiply this equation on the left by the "adjoint" matrix of A, we have

$$\left(\det\left(ba^{-1}\delta_{ij} - r_{ij}\right)\right) m_k = 0, \quad k = 1, \ldots, n.$$

(Recall that $\mathrm{adj}(A)$, the adjoint matrix of A, has the property $\mathrm{adj}(A)A = \det(A)I_n$, where I_n is the identity matrix.) Now the above equation implies

$$\det\left(ba^{-1}\delta_{ij} - r_{ij}\right) = 0$$

which in turn implies that ba^{-1} (which is not in R_P by Claim 1) satisfies a monic polynomial with coefficients in R_P, which is a contradiction to the hypothesis that R (and hence R_P) is normal and R_P is integrally closed. Hence our assumption is wrong, and Claim 2 follows.

For convenience, let $c = ba^{-1}$. By Claim 2, there exists a $d \in \mathfrak{m}$ such that $cd = 1 \in R_P$. Because $c \in \mathfrak{m}^{-1}$, for any $m \in \mathfrak{m}$ we have $cm \in R_P$. Thus $dcm = m$, proving that $\langle d\rangle = \mathfrak{m}$, i.e., the maximal ideal of R_P is principal. By Lemma 6.2.9, the maximal ideal of R_P has height 1, this in turn implies that $\mathrm{ht}(P) = 1$. This completes the proof that R satisfies (S_2) (in view of Lemma 6.2.8).

To show that R satisfies (R_1), we note that the discussion above shows that if P is the prime divisor of a principal ideal, then $\dim R_P = 1$, and the maximal ideal is generated by one element, and thus R_P is regular (cf. [28, Theorem 6.2A]). Since R is an integral domain, the only height 0 prime ideal is (0), in which case $R_{(0)}$ is a field and hence regular. The only ideals left to consider are prime ideals of height 1 that are not prime divisors of principal ideals. From the argument used to prove the converse of Lemma 6.2.8, such an ideal has depth at least 2. By Proposition 2.1.14, such a prime ideal has height at least 2. Thus R satisfies (R_1). □

Lemma 6.2.11. *Let R be Noetherian. For $b \in R$ and $x \in R \setminus \langle b \rangle$, define the ideal $I_x = \{r \in R \mid rx \in \langle b \rangle\}$. Then a maximal element of the set $\{I_x\}_{x \in R \setminus \langle b \rangle}$ is a prime ideal.*

Proof. Suppose I_s is a maximal element of the set above for some $s \in R$; and suppose $r_1 r_2 \in I_s$ where $r_1 \notin I_s$. Then $sr_1 r_2 \in \langle b \rangle$, but $sr_1 \notin \langle b \rangle$. This implies $I_s + \langle r_2 \rangle \subseteq I_{sr_1}$. Since I_s is maximal, we must have $r_2 \in I_s$ (and $I_s = I_{sr_1}$), implying I_s is prime. □

Theorem 6.2.12. *If R satisfies (R_1) and (S_2), then R is normal.*

Proof. Since normality is a condition on all localizations of R, and conditions (R_1) and (S_2) pass to localizations, we may suppose that R is local.

Suppose $\frac{a}{b} \in K(R)$ (where $K(R)$ is the full ring of fractions of R) such that

$$\left(\frac{a}{b}\right)^n + c_{n-1}\left(\frac{a}{b}\right)^{n-1} + \ldots + c_1\frac{a}{b} + c_0 = 0 \qquad (\dagger)$$

where $a, b, c_0, \ldots, c_{n-1} \in R$ and b is a nonzero divisor. We want to show that $\frac{a}{b} \in R$, which is equivalent to showing $a \in bR$. If b is a unit, the result is clear. Let then b be a nonunit (so that $\langle b \rangle$ is a proper ideal).

Let P be a prime divisor of $\langle b \rangle$. Although it remains to be shown that R is an integral domain, we can use the argument from the proof of Lemma 6.2.8 to say that $\mathrm{ht}(P) \leq 1$; and since $\langle b \rangle \subset P$, we conclude that $\mathrm{ht}(P) = 1$ (note that $\mathrm{ht}(P) = 0$ would imply that $PR_P = (0)$ which in turn would imply $bs = 0$, for some $s \in R \setminus P$, which is not true, since b is not a zero divisor). By condition (R_1), R_P is a regular local ring, and by Remark 2.1.24, R_P is normal. Let us now regard $(\dagger)$ as a monic polynomial equation over R_P satisfied by $\frac{a}{b}$. We have $\frac{a}{b} \in R_P$, and thus $a \in bR_P$.

For $x \in R \setminus \langle b \rangle$, define the ideal $I_x = \{r \in R \mid rx \in \langle b \rangle\}$. Let us assume, if possible, that $a \notin \langle b \rangle$ $(\subset R)$. Then, there exists some maximal element I_x such that $I_x \supseteq I_a$. By Lemma 6.2.11, I_x is prime, let us denote it P. By definition, P is a prime divisor of $\langle b \rangle$. As shown above, $a \in bR_P$; hence there exist $r, s \in R$, $s \notin P$ such that $a = b\frac{r}{s}$. Then $as = br$, which implies $s \in I_a \subset P$, a contradiction. Therefore our assumption (that $a \notin \langle b \rangle$) is wrong and, $a \in bR$. Thus we have shown that R contains all elements of $K(R)$ which are integral over R.

Now we show that R is an integral domain. Let S be the set of nonzero divisors in R, and denote $K(R) = S^{-1}R$ (the full ring of fractions of R). Since R is reduced, we have by Fact 6 (following Lemma 6.2.7) that $K(R) = K(R/P_1) \times \ldots \times K(R/P_n)$, where $\{P_i\}_{1 \leq i \leq n}$ are the minimal prime divisors of (0), and $K(R/P_i)$ is a field for each i. Let e_i be the multiplicative identity of $K(R/P_i)$. Then clearly $e_i^2 = 1$, and $e_i e_j = 0$ for $i \neq j$. Since the equation $x^2 - x = 0$ is satisfied by e_i, $1 \leq i \leq n$, the set $\{e_i\}_{1 \leq i \leq n}$ is integral over R and hence contained in R. It follows that $R = Re_1 \times \ldots \times Re_n$. Recall from the beginning of the proof, that R is a local ring, and hence has one maximal ideal, implying $n = 1$, and thus (0) is a prime ideal (note that $(0) = \bigcap_{i=1}^{n} P_i$). Therefore there are no nonzero zero divisors in R. Thus, R is an integral domain, completing the proof of the Theorem. □

Combining Theorems 6.2.10 and 6.2.12, we have shown Serre's criterion for normality:

Corollary 6.2.13 (Serre's criterion). *A Noetherian ring R is normal if and only if R satisfies (R_1) and (S_2).*

6.2.1 *Factoriality*

In this section, we prove some results on an unique factorization domain (abbreviated UFD) which will be used in the discussion on the unique factoriality of Schubert varieties. We first prove two lemmas concerning ideal intersections in a Noetherian integral domain. We will need to use the following version of Nakayama's Lemma (cf. [21, Cor 4.7]).

Nakayama's Lemma: Let A be a Noetherian ring, and M a finitely generated A-module. Let $\mathfrak{a}$ be an ideal in A such that $\mathfrak{a}M = M$. Then there exists an $a \in 1 + \mathfrak{a}$ which annihilates M.

Lemma 6.2.14. *Let A be a Noetherian integral domain, x a non-unit, and $I = Ax$. Then $\bigcap_{n \in \mathbb{N}} I^n = (0)$.*

Proof. Let $J = \bigcap_{n \in \mathbb{N}} I^n$. We claim that the inclusion $Jx \subseteq J$ is in fact an equality. To see this, since every $z \in J$ is divisible by x, we can write $J = J'x$, for a suitable ideal J'. Let $y \in J'$; then yx (being in J) is divisible by x^n, $\forall n \in \mathbb{N}$. This implies that y is divisible by x^n, $\forall\, n \in \mathbb{N}$, and hence $y \in J$. Hence $J' \subseteq J$. This implies that $J = J'x \subseteq Jx$, and the claim follows.

We now apply Nakayama's Lemma, with $\mathfrak{a} = I, M = J$ (note that $\mathfrak{a}M = M$, in view of claim above). By Nakayama's Lemma, there exists $a \in 1+I$, say $a = 1+bx$, such that $(1 + bx)J = 0$ which implies that $J = (0)$ (since A is a domain), and the result follows. □

More generally, we have the following lemma.

Lemma 6.2.15. *Let A be a Noetherian integral domain. Then $\bigcap_{n\in\mathbb{N}} I^n = 0$, for any ideal I in A.*

Proof. The ideal $I \subseteq \mathfrak{m}$, for some maximal ideal $\mathfrak{m}$. Then $I_{\mathfrak{m}}$ is an ideal in the local ring $(A_{\mathfrak{m}}, \mathfrak{m}A_{\mathfrak{m}})$; thus $I_{\mathfrak{m}} \subseteq \mathfrak{m}A_{\mathfrak{m}}$ (= the Jacobson radical of $A_{\mathfrak{m}}$ - recall that the *Jacobson radical* in a ring is the intersection of all maximal ideals). Hence, by the "Krull intersection theorem" (which states "In a Noetherian ring R, we have $\bigcap_{n\in\mathbb{N}} I^n = 0$, for all ideals I contained in the Jacobson-radical of R," see [72, Theorem 8.10] for proof), we obtain that $\bigcap_{n\in\mathbb{N}} I^n_{\mathfrak{m}} = 0$. The required result follows from this, since A is a domain we have an inclusion $\bigcap_{n\in\mathbb{N}} I^n \subseteq \bigcap_{n\in\mathbb{N}} I^n_{\mathfrak{m}}$. □

Lemma 6.2.16. *Let A be a Noetherian integral domain, and s a prime element in A. If A_s is a UFD, then so is A.*

(Here, $A_s = S^{-1}A$, with $S = \{1, s, s^2, \cdots\}$.)

Proof. Let p be a prime element in A_s. We may suppose that $p \in A$, and $s \not| p$.

Claim (1). p is a prime element in A.

Say, $p|ab$, $a, b \in A$. This implies $p|a$ or $p|b$ in A_s (since p is a prime element in A_s), say, $p|a$ in A_s. Then $a = p(\frac{c}{s^r})$, for some $c \in A$, $r \in \mathbb{N}$. This implies $s^r a = pc$, which in turn implies that $s|c$ (since s is a prime and $s \not| p$), say $c = sc_1$. Hence we obtain that $s^{r-1}a = pc_1$, where by the same reasoning, we have that $s|c_1$; this implies that $s^2|c$. Thus proceeding, we obtain that $s^r|c$, say, $c = s^r c'$, for some $c' \in A$; hence, substituting $c = s^r c'$ in $a = p(\frac{c}{s^r})$, we obtain that $a = pc'$, i.e., $p|a$ and Claim 1 follows.

Claim (2). Up to units, any element $a \in A, a \neq 0$ is a product of prime elements in A.

To prove Claim 2, we first observe that given $a \in A$, either $s \not| a$ or there exists a positive integer r such that $s^r|a$, $s^{r+1} \not| a$ (using Lemma 6.2.14). Thus any $a \in A$ can be written as $a = s^r a'$, $r \geq 0$, $a' \in A$, $s \not| a'$. Since A_s is a UFD, we can write $\frac{a'}{1}$ as a product of primes in A_s:

$$\frac{a'}{1} = \frac{p_1 \cdots p_n}{s^m},\ m \geq 0,$$

where $p_i \in A$, $s \not| p_i$, $1 \leq i \leq n$. Hence, by Claim 1, $\{p_i, 1 \leq i \leq n\}$ are primes in A. This implies that $s^m a' = p_1 \cdots p_n$. This together with the facts that s is a prime and $s \not| p_i$, $1 \leq i \leq n$ implies that $m = 0$, and $a' = p_1 \cdots p_n$. We obtain that a' (and therefore $a = s^r a'$) is a product of primes in A, proving Claim 2.

Note that by Remark 2.1.23, Claim 2 implies that A is a UFD, as required. □

Remark 6.2.17. Lemma 6.2.16 may be thought of as a kind of converse to the result that if A is a UFD, then so is $S^{-1}A$, for any multiplicatively closed set S in A.

We include an alternate proof of Lemma 6.2.16 which uses the following characterization of unique factorization property for Noetherian domains.

Theorem 6.2.18. *Let A be a Noetherian integral domain. Then A is a UFD if and only if every height* 1 *prime is principal.*

Proof. $\Rightarrow$: Let $\mathfrak{p}$ be a height 1 prime. Let $s \in \mathfrak{p}$. Then $As \subseteq \mathfrak{p}$; we can write $s = p_1 \cdots p_n$, p_i prime (since A is a UFD). This implies that $p_i \in \mathfrak{p}$, for some i. Denote $p = p_i$. Then the inclusion $Ap \subseteq \mathfrak{p}$ is in fact an equality (since $\mathfrak{p}$ is a height 1 prime and A is a domain), proving that $\mathfrak{p}$ is principal.

$\Leftarrow$: Let $x \in A$. We can write $x = x_1 \cdots x_n$, x_i irreducible (since A is Noetherian). We shall show that under the hypothesis that every height 1 prime is principal, an irreducible element $a \in A$ is in fact prime. Using "Krull's principal ideal theorem," which states "In a Noetherian ring, every principal ideal has height ≤ 1" where the height of a general ideal I is defined to be the infimum of the heights of prime ideals containing I (cf. [21, §8.2.2]), we get that any minimal prime $\mathfrak{p}$ containing a has height ≤ 1. In fact, the height of $\mathfrak{p}$ equals 1, since A is a domain. Hence, the hypothesis implies that such a prime ideal $\mathfrak{p}$ is in fact principal, say $\mathfrak{p} = Ap$ where p is a prime element. We have, $Aa \subseteq \mathfrak{p}$ (since $a \in \mathfrak{p}$). Hence, $a = bp$, for some $b \in A$; in fact, b is a unit (since a is irreducible). This implies that a is prime, as asserted. □

Proof (An alternate proof of Lemma 6.2.16). Let $\mathfrak{p}$ be a height 1 prime ideal in A. Under the hypothesis that A_s is a UFD and s is a prime element, we shall show that $\mathfrak{p}$ is principal. If $s \in \mathfrak{p}$, then $As \subseteq \mathfrak{p}$; hence, $\mathfrak{p} = As$ (since height of $\mathfrak{p}$ is 1 and A is a domain). Let then $s \notin \mathfrak{p}$. Then $\mathfrak{p}_s$ is a height 1 prime ideal in A_s, and therefore principal (in view of the hypothesis that A_s is a UFD), say,

$$\mathfrak{p}_s = A_s p, \tag{*}$$

p being a prime element in A_s. We may suppose that $p \in A$ and $s \not| p$. Now (*) implies that

$$s^m \mathfrak{p} \subseteq Ap \tag{**}$$

for some positive integer m. Let $x \in \mathfrak{p}$; then by (**), $s^m x = ap$. Now the facts that s is a prime and that $s \not| p$ imply that $s^m | a$, say, $a = s^m a'$; substituting in $s^m x = ap$, we obtain $x = a'p$, and hence $\mathfrak{p} = Ap$, as required. □

We conclude this section with one interesting property on UFD's.

Theorem 6.2.19. *Let A be a Noetherian graded normal domain with $A_0 = K$, a field. Let $\mathfrak{m} = A_+$. Suppose $A_\mathfrak{m}$ is an UFD. Then A is an UFD.*

Proof. We shall use the following characterization of the unique factorization property for graded Noetherian domains (similar to Theorem 6.2.18):

(†) *Let A be a Noetherian graded integral domain. Then A is a UFD if and only if every height* 1 *graded prime ideal is principal.*

Let $\mathfrak{p}$ be a height 1 graded prime ideal in A. Then $\mathfrak{p}_\mathfrak{m}$ is a height 1 prime ideal in $A_\mathfrak{m}$. The hypothesis that $A_\mathfrak{m}$ is an UFD implies (in view of Theorem 6.2.18) that $\mathfrak{p}_\mathfrak{m}$ is a principal ideal in $A_\mathfrak{m}$, say,

$$\mathfrak{p}_\mathfrak{m} = A_\mathfrak{m} x \tag{*}$$

for some $x \in A$. Let z be a homogeneous element in $\mathfrak{p}$. Then in view of (*), we can write

$$z = \frac{y}{s} x \tag{**}$$

for some $y \in A$, $s \in A \setminus \mathfrak{m}$. Since, $s \in A \setminus \mathfrak{m}$, s has the form $s = s_0 +$ (higher degree terms), where $s_0 \in K^*$, while $x = x_{m_1} + x_{m_2} + \cdots$, where $x_{m_i} \in \mathfrak{p}$ for all i, and $x_{m_1} \neq 0$ (here, the subscripts denote the respective degrees of the homogeneous components of x).

Let $y = y_{k_1} + y_{k_2} + \cdots$, where $y_{k_1} \neq 0$. Substituting for x, y, s in (**), and comparing the lowest degree terms, we get $zs_0 = y_{k_1} x_{m_1}$ (where note that s_0 is a unit). This implies that $\mathfrak{p} \subseteq Ax_{m_1} \subseteq \mathfrak{p}$, i.e., $\mathfrak{p} = Ax_{m_1}$ is principal, as required. □

6.3 Normality

In this section, we show that Schubert varieties are arithmetically normal. We first show that they are normal using Serre's criterion, which includes a discussion of the regularity in codimension one for Schubert varieties (see Definition 6.3.1 below). It is a result due to Chevalley that Schubert varieties are regular in codimension one.

Definition 6.3.1. An algebraic variety is said to be *regular in codimension one* if its singular locus has codimension at least two.

We give below the definition of a "reflection," which applies only to $SL(n)$ and does not require abstract knowledge of root systems.

Definition 6.3.2. Let s be an element of the symmetric group S_n (the group of permutations of $\{1, \ldots, n\}$); s is a *simple reflection* if s is a transposition of two adjacent numbers, i.e., $s = (i, i+1)$ for $1 \leq i \leq n-1$. More generally, the transposition of i, j, $1 \leq i, j \leq n$ is called a *reflection*.

A reflection acts on $I_{d,n}$ as follows: for $s = (i, j)$ and $\tau \in I_{d,n}$, if precisely one of $\{i, j\}$ appears in τ, then $s\tau$ is obtained by replacing one with the other in τ (and then all indices are re-ordered to be increasing). If neither or both of $\{i, j\}$ appear in τ, then $s\tau$ is just τ.

Remark 6.3.3. If $X_{w'}$ is a Schubert divisor of $X_w (\subset G_{d,n})$, then there exists a simple reflection s such that $w' = sw$; if $w = (i_1, \ldots, i_d)$, then there exists t such that $w' = (i_1, \ldots, i_t - 1, \ldots i_d)$. Then $w' = sw$, where $s = (i_t - 1, i_t)$.

6.3.1 Stability for multiplication by certain parabolic subgroups

For i, $1 \leq i \leq n-1$, let U_{-i} be the one-dimensional subgroup in G consisting of all lower triangular unipotent matrices with a possible nonzero entry only at the $(i+1, i)^{th}$ place, and let $\mathcal{P}_i$ be the parabolic subgroup generated by B, U_{-i}. For $w \in S_n$, let

$$\mathcal{X}(w) = \overline{BwB} \subset G,$$

here, by BwB, we mean Bn_wB, n_w being a lift for w in $N(T)$ (we identify S_n with $N(T)/T$, T being the maximal torus of diagonal matrices in G and $N(T)$={permutation matrices}, matrices such that in each row and column, there is a unique nonzero entry, $N(T)$ is the normalizer of T in G). Note that the set BwB is independent of the lift n_w.

We are interested in finding a characterization of w's such that $\mathcal{X}(w)$ is stable for multiplication on the right (resp. left) by $\mathcal{P}_i$, $1 \leq i \leq n-1$. Let π_r (resp. π_l) be the canonical map $G \to G/B$ (respectively, $G \to B \setminus G$). Let us denote $\pi_r(\mathcal{X}(w))$ (resp. $\pi_l(\mathcal{X}(w))$) by $\mathcal{X}(w)_r$ (resp. $\mathcal{X}(w)_l$).

Lemma 6.3.4. *With notation as above, $\mathcal{X}(w)$ is stable for multiplication on the right (resp. left) by $\mathcal{P}_i$, $1 \leq i \leq n-1$, if and only if $\mathcal{X}(w)_r \supset \mathcal{X}(ws_i)_r$ (resp. $\mathcal{X}(w)_l) \supset \mathcal{X}(s_iw)_l$)). Here, s_i is the transposition $(i, i+1)$.*

Proof. Let $\theta_{i,r}$ (resp. $\theta_{i,l}$) be the canonical map $G/B \to G/\mathcal{P}_i$ (resp. $B\backslash G \to \mathcal{P}_i\backslash G$). Then the fibers of $\theta_{i,r}$ (resp. $\theta_{i,l}$) are isomorphic to $\mathcal{P}_i/B$ (resp. $B\backslash\mathcal{P}_i$). Thus $\theta_{i,r}$ (resp. $\theta_{i,l}$) is a $\mathbb{P}^1$-fibration. Also, for $w \in S_n$, we have

$$(*) \qquad \theta_{i,r}(\mathcal{X}(w)_r) = \theta_{i,r}(\mathcal{X}(ws_i)_r), \; \theta_{i,l}(\mathcal{X}(w)_l) = \theta_{i,l}(\mathcal{X}(s_iw)_l).$$

Clearly, $\mathcal{X}(w)$ is stable for multiplication on the right (resp. left) by $\mathcal{P}_i$ if and only if $\mathcal{X}(w) = \overline{Bw\mathcal{P}_i}$ (resp. $\overline{\mathcal{P}_iBw}$); we have that $\mathcal{X}(w) = \overline{Bw\mathcal{P}_i}$ (resp. $\overline{\mathcal{P}_iBw}$) if and only if $\mathcal{X}(w)_r$ (resp. $\mathcal{X}(w)_l$) is saturated for the $\mathbb{P}^1$-fibration $\theta_{i,r}$ (resp. $\theta_{i,l}$), i.e., $\mathcal{X}(w)_r = \theta_{i,r}^{-1}(\theta_{i,r}(\mathcal{X}(w)_r))$ (resp. $\mathcal{X}(w)_l = \theta_{i,l}^{-1}(\theta_{i,l}(\mathcal{X}(w)_l))$); this together with $(*)$ implies the result. □

Before moving on, we cite the following Lemma from [6, 1.8], known as the "closed orbit lemma."

Lemma 6.3.5 (Closed orbit lemma). *Let X be an algebraic variety together with an action by a connected algebraic group G. Let $x \in X$.*

1. *The orbit Gx is open in $\overline{Gx}$.*
2. *$\overline{Gx} \setminus Gx$ is a union of orbits of strictly smaller dimension. In particular, orbits of minimal dimension are closed (and thus closed orbits exist).*

Recall the notation from Chapter 5: for $w = (i_1, \ldots, i_d) \in I_{d,n}$, $[e_w] = [e_{i_1} \wedge \cdots \wedge e_{i_d}] \in \mathbb{P}\left(\bigwedge^d K^n\right)$.

Lemma 6.3.6. *The point $[e_w]$ is a smooth point of X_w.*

Proof. By (1) of the Closed orbit lemma, we have that the B-orbit $B \cdot [e_w]$ is an open subset of X_w. Hence, the point $[e_w]$ is a smooth point of X_w if and only if it is a smooth point of the orbit $B \cdot [e_w]$. From Remark 5.3.9, we have a canonical identification of $B \cdot [e_w]$ with $\mathbb{A}^{\dim X_w}$; hence, $[e_w]$ is a smooth point of $B \cdot [e_w]$, and the required result follows. □

Lemma 6.3.7. *Let $X_{w'}$ be a Schubert divisor in $X_w (\subset G_{d,n})$, $w, w' \in I_{d,n}$. The point $[e_{w'}]$ is a smooth point of X_w.*

Proof. Let $w = (i_1, \ldots, i_d)$. By Remark 6.3.3, $w' = sw$, for some simple reflection s; say $s = (i, i+1)$. Identifying w with the permutation $(i_1 \ldots i_d j_1 \ldots j_{n-d})$, ($\{j_1 \ldots j_{n-d}\}$ being the complement of $\{i_1, \ldots, i_d\}$, in $\{1, \ldots, n\}$), we have that $sw < w$, and hence by Lemma 6.3.4, we obtain that $\mathcal{X}(w)$ is stable for multiplication on the left by the parabolic subgroup $\mathcal{P}_i$. This implies (by considering the canonical projection $G \to G/P_d (= G_{d,n})$) that $X(w) (\subset G_{d,n})$ is stable for multiplication on the left by the parabolic subgroup $\mathcal{P}_i$. In particular, taking a lift n_s for s in $N(T)$ (note that $n_s \in \mathcal{P}_i$), we have that $X(w)$ is stable for multiplication by n_s (on the left), and thus left multiplication by n_s induces an automorphism of $X(w)$, under which, clearly e_w gets mapped to $e_{w'}$. The result now follows from Lemma 6.3.6. □

Combining Lemmas 6.3.6, 6.3.7, we obtain the following.

Theorem 6.3.8. *For any $w \in I_{d,n}$, the Schubert variety $X(w)$ in $G_{d,n}$ is regular in codimension 1.*

Theorem 6.3.9. *The Schubert variety X_τ in $G_{d,n}$ is arithmetically normal; i.e., the cone over X_τ is normal.*

Proof. We have that X_τ is Cohen–Macaulay, and hence the stalks $\mathcal{O}_{X_\tau,x}$, $x \in X_\tau$ are Cohen–Macaulay; in particular, $\mathcal{O}_{X_\tau,x}$ satisfies (S_2). We also have that X_τ is regular in codimension 1, which is equivalent to the stalks $\mathcal{O}_{X_\tau,x}$, $x \in X_\tau$ satisfying (R_1). Therefore, by Serre's criterion for normality, the stalks $\mathcal{O}_{X_\tau,x}$, $x \in X_\tau$ are normal, and hence X_τ is normal.

It remains to be shown that X_τ is arithmetically normal. We note that if a projective variety $X \hookrightarrow \mathbb{P}^n$ is normal, then it is arithmetically normal if and only if the restriction map $H^0(\mathbb{P}^n, L^m) \to H^0(X, L^m)$ is surjective for all $m \in \mathbb{Z}_+$, where $L = \mathcal{O}_{\mathbb{P}^n}(1)$ (cf. [28, Ex. 5.14 (d)]). Let ϕ_m be this restriction map:

$$\phi_m : H^0(Y, L^m) \to H^0(X_\tau, L^m),$$

where $Y = \mathbb{P}\left(\bigwedge^d K^n\right)$. Now, the image of ϕ_m is $R(\tau)_m$ ($R(\tau) = K[X_\tau]$, the homogeneous coordinate ring of X_τ for the Plücker embedding). By Corollary 5.6.5, this is exactly $H^0(X_\tau, L^m)$. Thus the map ϕ_m is surjective for all m, and thus X_τ is arithmetically normal. □

6.4 Factoriality

Now that we have shown that all Schubert varieties in $G_{d,n}$ are arithmetically Cohen–Macaulay and arithmetically normal, in this section we will see that not all Schubert varieties are factorial. We first introduce the "opposite big cell in $G_{d,n}$" and the "opposite cell in a Schubert variety."

Definition 6.4.1. Define U_d^- as:

$$U_d^- = \left\{ \begin{pmatrix} I_d & 0_{d\,n-d} \\ X_{n-d\,d} & I_{n-d} \end{pmatrix} \right\}$$

$X_{n-d\,d}$ being an $(n-d) \times d$ matrix. Then under the canonical projection $\pi_d : G \to G/P_d$, U_d^- is mapped isomorphically onto its image $\left\{ \begin{pmatrix} I_d \\ X_{n-d\,d} \end{pmatrix} \right\}$. Further, $\pi_d(U_d^-)$ is simply $B^- e_{id}$, where B^- is the subgroup of G consisting of lower triangular matrices (in the literature, B^- is called *the Borel subgroup opposite to* B); $B^- e_{id}$ is called *the opposite big cell* in $G_{d,n}$, and is denoted $\mathcal{O}_d^-$. This gives an identification $\mathcal{O}_d^- \cong M_{n-d\,d}$ (the space of $(n-d) \times d$ matrices with entries in K). Thus we obtain that $B^- e_{id} \cong \mathbb{A}^{d(n-d)}$. We also have an identification of $B^- e_{id}$ with $\{a \in G_{d,n} \mid p_{(1,\ldots,d)}(a) \neq 0\}$.

Given $w = (i_1, \ldots, i_d) \in I_{d,n}$, define $Y(w) := X_w \cap B^- e_{id}$; note that $Y(w)$ sits as a closed subvariety of $\mathbb{A}^{d(n-d)}$. Thus $Y(w)$ is an affine open subset in X_w, called the *opposite cell* in X_w (by abuse of terminology; note that $Y(w)$ need not be a cell in the topological, since it need not be smooth). Also, we have,

$$Y(w) = \{a \in X_w \mid p_{(1,\ldots,d)}(a) \neq 0\}.$$

Theorem 6.4.2. *Let X_τ be a Schubert variety in $G_{d,n}$. Then the following statements are equivalent:*

1. *X_τ is arithmetically factorial, i.e., $R(\tau)$ (the homogeneous coordinate ring of X_τ for the Plücker embedding) is an UFD.*
2. *The element p_τ is prime in $R(\tau)$.*
3. *X_τ has one unique Schubert divisor.*
4. *Either $\tau = (i_1, \ldots, i_d)$ consists of one segment of consecutive integers, or $i_1 = 1$ and τ consists of two segments of consecutive integers (e.g., $(1, 2, 6, 7, 8) \in I_{5,9}$).*
5. *X_τ is isomorphic to a suitable Grassmannian variety.*
6. *X_τ is nonsingular.*
7. *X_τ is factorial (i.e., the stalks $\mathcal{O}_{X_\tau,x}$, $x \in X_\tau$ are UFD's).*
8. *$Y(\tau)$ is factorial.*

Proof. Before we begin showing the implications above, we note that p_w, $w \leq \tau$ is an irreducible element in $R(\tau)$ (i.e., the ideal $\langle p_w \rangle$ is maximal among the principal ideals in $R(\tau)$), since $R(\tau)$ is a graded ring generated by the degree 1 elements p_w, $w \leq \tau$. Thus (1)⇒(2) by Remark 2.1.23.

Claim. (2)⇒(3): Let p_τ be prime in $R(\tau)$, and let $X_1, \ldots, X_r$ be all of the Schubert divisors of X_τ. Let I_j be the ideal defining X_j in X_τ, $1 \leq j \leq r$. Since Schubert divisors are of codimension 1 in X_τ, each I_j is a prime ideal with height 1. By Pieri's formula (5.5.8), $\langle p_\tau \rangle = \bigcap_{j=1}^r I_j$; hence $r = 1$ and $\langle p_\tau \rangle = I_1$.

Claim. (3)⇒(4): Assume if possible that τ contains at least two segments of consecutive integers, say $(i_r, i_r + 1 \ldots, i_{s-1}), (i_s, i_s + 1 \ldots, i_d)$, $i_r \notin \{1, i_{r-1} + 1\}$, and $i_{s-1} + 1 \neq i_s$ (if $r = 1$, then i_{r-1} is understood to be 0). Given these conditions, we may define $\underline{a}, \underline{b} \in I_{d,n}$ as such:

$$a_j = \begin{Bmatrix} i_j & \text{if } j \neq r \\ i_r - 1 & \text{if } j = r \end{Bmatrix}, \quad b_j = \begin{Bmatrix} i_j & \text{if } j \neq s \\ i_s - 1 & \text{if } j = s \end{Bmatrix}.$$

Now $X_{\underline{a}}$ and $X_{\underline{b}}$ are distinct Schubert divisors in X_τ, a contradiction.

Claim. (4)⇒(5): Note that if τ consists of one segment, then X_τ is isomorphic to the largest Schubert variety in G_{d,i_d}, which of course is the Grassmannian G_{d,i_d}. On the other hand, if $\tau = (1, \ldots, r-1, i_r, \ldots, i_d)$, then X_τ is isomorphic to $X_{(i_r,\ldots,i_d)} \subseteq G_{d-r+1,n}$, and we have $(i_r, \ldots, i_d)$ consisting of one segment again.

The implication (5)⇒(6) follows from Corollary 6.5.3 (cf. the discussion in the next section on the singular loci of Schubert varieties); (6)⇒(7) follows from Remark 2.1.24,(4); and (7)⇒(8) is clearly true, since $Y(\tau)$ is an affine open subset of X_τ.

The assertion that (8)⇒(1) follows from Proposition 6.4.5 proved below. □

We shall first prove some preliminary results leading to the proof of the assertion that $Y(w)$ is factorial implies that X_w is arithmetically factorial (i.e., $R(w)$ the homogeneous coordinate ring of X_w for the Plücker embedding is an UFD).

Let R be a graded domain, $R = \bigoplus_{i \geq 0} R_i$; and s a homogeneous element of degree $d > 0$. Let S be the homogeneous localization of R with respect to s; note that S consists of $\frac{f}{s^t}$, $f \in R_{td}$.

Lemma 6.4.3. *With notation as above, s is a transcendental over S.*

Proof. The proof is rather immediate. If possible, let us assume that s satisfies a polynomial equation over S:

$$F_n s^n + \cdots + F_i s^i + \cdots + F_1 s + F_0 = 0,$$

$F_i \in S$, say $F_i = \frac{f_i}{s^{r_i}}$, $f_i \in R_{dr_i}$, $0 \leq i \leq n$, and $f_n \neq 0$. Let $M = \{1, s, s^2, \ldots\}$. For the $\mathbb{Z}$-grading of $M^{-1}R$ (induced from the grading of R), degree of $F_i s^i = di$. Hence, homogeneity implies that $F_i s^i = 0$, for all $0 \leq i \leq n$; in particular, $F_n = 0$, a contradiction. Thus our assumption is wrong and the result follows. □

The next result relates a given Schubert variety X_w and a certain divisor in $G_{d,n}$. Let B^- be as in Definition 6.4.1. We consider the B^--orbits in $G_{d,n}$. The B^--orbit closures are again indexed by $I_{d,n}$, and for $w \in I_{d,n}$, the B^--orbit closure

$\overline{B^- e_w}$ is called the *opposite Schubert variety associated to* w, and is denoted $X^-(w)$. It is easily seen that for $w, \tau \in I_{d,n}$, $X^-(w) \subset X^-(\tau)$ if and only if $w > \tau$. In particular, $G_{d,n} = X^-((1,2,\ldots,d))$, and the unique opposite Schubert divisor of $G_{d,n}$ (i.e., the B^--orbit closure of codimension 1 in $G_{d,n}$) is given by $X^-(w), w = (1,2,\ldots,d-1,d+1)$. Proceeding as in Theorem 5.5.8, we have that the intersection $G_{d,n} \cap \{p_{(1,2,\ldots,d)} = 0\}$ is scheme-theoretic, and equals $X^-((1,2,\ldots,d-1,d+1))$; further, $K[X^-(1,2,\ldots,d-1,d+1)]$, the homogeneous coordinate ring of $X^-((1,2,\ldots,d-1,d+1))$, has a basis consisting of standard monomials not involving $p_{(1,\ldots,d)}$. Similarly, $Z^-(w) := X_w \cap \{p_{(1,2,\ldots,d-1,d)} = 0\} (= X_w \cap X^-((1,2,\ldots,d-1,d+1)))$ is again scheme-theoretic (and reduced, irreducible); further, $K[Z^-(w)]$ has a basis consisting of monomials standard on X_w not involving $p_{(1,\ldots,d)}$. Hence we obtain the following.

Proposition 6.4.4. *The kernel of the (surjective) restriction map* $K[X_w] \to K[Z^-(w)]$ *is precisely the principal ideal generated by* $p_{(1,\ldots,d)}$. *In particular,* $p_{(1,\ldots,d)}$ *is a prime element in* $K[X_w]$.

Denote $R(w)$ by R, and the Plücker coordinate $p_{(1,\ldots,d)}$ by s. Denote the homogeneous localization of R with respect to s by S. Let $Y(w)$ be the opposite cell in X_w, namely, $Y(w) = X_w \cap B^- e_{id}$. Then $Y(w) = Spec\, S$.

Proposition 6.4.5. *Let* $Y(w)$ *be factorial (namely,* S *is an UFD). Then* $X(w)$ *is arithmetically factorial, i.e.,* R *is an UFD.*

Proof. The hypothesis together with Lemma 6.4.3 implies that $S[s]$ is an UFD; hence $S[s]_s (= S[s,s^{-1}])$ is an UFD. This implies that $R_s (= S[s,s^{-1}])$ is an UFD. Hence, we obtain (in view of Proposition 6.4.4 and Lemma 6.2.16) that R is an UFD. □

This completes the proof of Theorem 6.4.2.

Corollary 6.4.6. *A Schubert variety is factorial if and only if it is arithmetically factorial if and only if it is nonsingular.*

Example 6.4.7. Let us list all factorial (and hence nonsingular) Schubert varieties in $G_{3,6}$. All 20 Schubert varieties of $G_{3,6}$ are listed in Example 5.3.3. Using statements (3) or (4) of Theorem 6.4.2, it is easy to check that the following ten Schubert varieties are the only ones that are factorial, nonsingular:

$$X_{(123)}, X_{(124)}, X_{(125)}, X_{(126)}, X_{(134)}$$

$$X_{(145)}, X_{(156)}, X_{(234)}, X_{(345)}, \text{ and } X_{(456)} = G_{3,6}$$

6.5 Singular Locus

In this section, we shall determine the singular loci of Schubert varieties in the Grassmannian.

Recall from §2.8 that given a variety X, a point $x \in X$ is a smooth point if and only if $\dim T_xX = \dim X$. A Schubert variety $X_w \subseteq G_{d,n}$ is a union of B-orbits, and points within an orbit have isomorphic tangent spaces. Thus, for the discussion of the singular locus of X_w, we can restrict our attention to the set of T-fixed points $\{e_\tau, \tau \leq w\}$.

Let G be an affine algebraic group, $A = K[G]$, and let e denote the identity element of G; Lie G, the Lie algebra of G, is defined as follows.

Definition 6.5.1.

$$\text{Lie}\, G = \{D \in \text{Der}_K(A, A) \mid \lambda_x \circ D = D \circ \lambda_x, \text{ for all } x \in G\},$$

where $\lambda_x(f)(g) = f(x^{-1}g)$ for $f \in A$, $g \in G$; such a D is called a *left invariant vector field.*

We have a canonical identification: $T_eG = \text{Lie}\, G$ (see [6] for details). We denote Lie G by $\mathfrak{g}$.

Let $G = SL_n$. It is a well-known fact (cf. [6, §3.9]) that $\mathfrak{g}$ is the set of all traceless $n \times n$ matrices (i.e., matrices (m_{ij}) such that the sum of diagonal entries equals zero: $m_{11} + m_{22} + \cdots + m_{nn} = 0$). Letting P_d be the parabolic subgroup as described in §5.1, LieP_d is a Lie subalgebra of $\mathfrak{g}$; specifically, LieP_d consists of traceless matrices (m_{ij}) such that $m_{ij} = 0$ for $d+1 \leq i \leq n$ and $1 \leq j \leq d$.

This leads to an identification of the tangent space of G/P_d at eP_d with

$$\bigoplus_{1 \leq j \leq d < i \leq n} \mathfrak{g}_{ij},$$

where $\mathfrak{g}_{ij}$ is the one-dimensional K-span of the elementary matrix E_{ij}. We note that under the identification $G/P_d \cong G_{d,n}$, eP_d corresponds to the point $e_{id} (= e_{(1,\ldots,d)}) \in G_{d,n}$.

Thus we have

$$T_{e_{id}} G_{d,n} = \bigoplus_{1 \leq j \leq d < i \leq n} \mathfrak{g}_{ij}, \ \dim T_e G_{d,n} = d(n-d) = \dim G_{d,n}$$

Thus e_{id} is a smooth point of $G_{d,n}$.

Lemma 6.5.2. *The Schubert variety X_w is nonsingular if and only if X_w is smooth at the point e_{id}.*

Proof. The first direction is clear: if X_w is nonsingular, it is nonsingular at each point.

Suppose Sing X_w is nonempty. Then, because the singular locus is a closed, B-stable subset of X_w, the singular locus is a union of Schubert subvarieties of X_w. Thus $X_{e_{id}} \subseteq \text{Sing}\, X_w$, and the result follows. □

Corollary 6.5.3. *The Grassmannian variety $G_{d,n}$ is nonsingular.*

Remark 6.5.4. For the rest of the section, we will be using the "Jacobian criterion for smoothness:" Let Y be an affine variety in $\mathbb{A}^n$, and let $I(Y)$ be the ideal in $K[x_1, \ldots, x_n]$ defining Y. Let $I(Y)$ be generated by $\{f_1, f_2, \ldots, f_r\}$. Let J be the Jacobian matrix $(\frac{\partial f_i}{\partial x_j})$. For $P \in Y$, we have, rank $J_P \leq \mathrm{codim}_{\mathbb{A}^n} Y$ with equality if and only if P is a smooth point of Y (here J_P denotes J evaluated at P).

Consider the (open) subset $B^- e_{id} (= \{x \in G_{d,n} | p_{id}(x) \neq 0\})$ in $G_{d,n}$. As seen in Definition 6.4.1, we have an identification of $B^- e_{id}$ with the set of $n \times d$ matrices of the form $\begin{bmatrix} Id_d \\ X_{(n-d)\times d} \end{bmatrix}$; hence, we shall identify $B^- e_{id}$ with $\mathbb{A}^{d(n-d)}$, and denote the affine coordinates on $\mathbb{A}^{d(n-d)}$ by $\{x_{ij}, 1 \leq j \leq d, d+1 \leq i \leq n\}$; clearly, e_{id} is identified with the origin in $\mathbb{A}^{d(n-d)}$.

For $w \in I_{d,n}$, consider $Y(w) \hookrightarrow B^- e_{id}$ ($\cong \mathbb{A}^{d(n-d)}$, cf. Definition 6.4.1). Clearly, $Y(w)$ is an affine neighborhood of e_{id} (in X_w), and hence, e_{id} is a smooth point of X_w if and only if it is a smooth point of $Y(w)$. Denoting $p_\theta|_{B^- e_{id}}$ by $f_\theta, \theta \in I_{d,n}$, we have by Theorem 5.4.8 that the ideal of $Y(w)$ (as a subvariety of $\mathbb{A}^{d(n-d)}$) is generated by $\{f_\theta, \theta \not\leq w\}$. Let J_w (or just J) be the Jacobian matrix of $Y(w)$. Note that the rows of J are indexed by $\{\theta \in I_{d,n} | \theta \not\leq w\}$, while the columns are indexed by $\{ij | 1 \leq j \leq d < i \leq n\}$; thus, we shall write $J = (a_{\theta,ij})$ (where $a_{\theta,ij} = \frac{\partial f_\theta}{\partial x_{ij}}, \theta, ij$ being as above). We shall now determine the smoothness/nonsmoothness of $Y(w)$ at e_{id} using the Jacobian criterion. Towards computing the rank of $J_{e_{id}}$ ($J_{e_{id}}$ being J evaluated at e_{id}), we shall first compute $\frac{\partial f_\theta}{\partial x_{ij}}$, where $\theta \not\leq w, 1 \leq j \leq d < i \leq n$.

Let $L = \mathcal{O}_{G_{d,n}}(1)$ as defined in §2.7. Since $H^0(G_{d,n}, L)$ is a G-module, it is also a $\mathfrak{g}$-module. Given X in $\mathfrak{g}$, we identify X with the corresponding left invariant vector field D_X on G. For $f \in H^0(G_{d,n}, L)$, we have $D_X f = Xf$; further, for $f \in H^0(G_{d,n}, L)$, denoting $f|_{B^- e_{id}}$ also by just f, we have that the evaluations of $\frac{\partial f}{\partial x_{ij}}$ and $X_{ij} \cdot f$ at e_{id} coincide (note that $X_{ij} \in \mathfrak{g}$ is just the elementary matrix $E_{ij}, 1 \leq j \leq d < i \leq n$). Thus we have reduced the problem to computing $X_{ij} \cdot f_\theta$, $\theta \not\leq w$, $1 \leq j \leq d < i \leq n$.

We have $G_{d,n} \hookrightarrow \mathbb{P}\left(\bigwedge^d V\right)$. Thus

$$H^0(G_{d,n}, L) = \left(\bigwedge{}^d V\right)^* = \bigwedge{}^{(n-d)} V, \quad V = K^n.$$

Let $\theta \in I_{d,n}$, say $\theta = (i_1, \ldots, i_d)$. Let $\{j_1, \ldots, j_r\}$ be the complement of θ in $\{1, \ldots, n\}$ (arranged in ascending order) where $r = n - d$. We have $e_{i_1} \wedge \ldots \wedge e_{i_d} \in \bigwedge^d V$. Therefore $p_\theta = e_{j_1} \wedge \ldots \wedge e_{j_r}$.

We have that $E_{ij} \cdot e_k = \delta_{kj} e_i$; and therefore

$$\begin{aligned} X_{ij} \cdot p_\theta &= \sum_t e_{j_1} \wedge \ldots \wedge E_{ij} e_{j_t} \wedge \ldots \wedge e_{j_r} \\ &= \begin{cases} 0, & \text{if } j \notin \{j_1, \ldots, j_r\} \text{ or } i, j \in \{j_1, \ldots, j_r\} \\ \pm p_{\theta'}, & \text{where } \theta' = \left(i_1, \ldots, \hat{i}, \ldots, i_d, j\right) \uparrow . \end{cases} \end{aligned}$$

We have, $\theta' = s\theta$ where s is the reflection (i,j) acting on $I_{d,n}$ as described in §6.3 (in the paragraph following Definition 6.3.2), and hence $X_{ij}f_\theta = \pm f_{\theta'}$.

Evaluating J at e_{id}, we have, for $1 \leq j \leq d < i \leq n$

$$\begin{aligned}(X_{ij}f_\theta)_{e_{id}} \neq 0 &\Leftrightarrow f_{\theta'} = c \cdot f_{id},\ c \in K^* \\ &\Leftrightarrow \theta' = (1, \ldots, d) \\ &\Leftrightarrow \theta = (1, \ldots, \hat{j}, \ldots, d, i).\end{aligned}$$

Thus in $J_{e_{id}} = (a_{\theta,ij})$, we have $a_{\theta,ij} \neq 0$ if and only if $\theta = (1, \ldots, \hat{j}, \ldots, d, i)$. Thus the only relevant rows are f_θ, $\theta = (1, \ldots, \hat{j}, \ldots, d, i)$ (for some (i,j), $1 \leq j \leq d < i \leq n$) such that $\theta \not\leq w$; further, in such a row there is precisely one nonzero entry (note that x_{ij} is the only variable such that $\frac{\partial}{\partial x_{ij}}(f_\theta)$ is nonzero).

Therefore, the rank of $J_{e_{id}}$ is equal to the number of nonzero columns, and this number is equal to the number of reflections (i,j) such that $w \ngeq (1, \ldots, \hat{j}, \ldots, d, i)$. Thus we obtain the following.

Theorem 6.5.5.

1. *The rank of $J_{e_{id}}$ is equal to $\#\{(i,j), 1 \leq j \leq d < i \leq n \mid w \ngeq (1, \ldots, \hat{j}, \ldots, d, i)\}$.*
2. $\dim T_{e_{id}}X_w = \#\{(i,j), 1 \leq j \leq d < i \leq n \mid w \geq (1, \ldots, \hat{j}, \ldots, d, i)\}$.

As a result of Lemma 6.5.2 and Remark 6.5.4, the rank of $J_{e_{id}} = d(n-d) - \dim X_w$ if and only if X_w is smooth. There are $d(n-d)$ reflections (i,j) such that $1 \leq i \leq d < j \leq n$. This gives us the following criteria for X_w to be smooth.

Corollary 6.5.6. *The Schubert variety X_w is nonsingular if and only if*

$$\dim X_w = \#\{(i,j) \mid 1 \leq j \leq d < i \leq n, \text{ and } w \geq (1, \ldots, \hat{j}, \ldots, d, i)\}.$$

Recall that $\dim X_w = \sum_{t=1}^{d} (i_t - t)$ for $w = (i_1, \ldots, i_d)$.

The discussion at any other $e_\tau \in X(w)$ is similar. Let $\tau = (a_1 \ldots a_d)$. One works with the affine neighborhood $Y(w,\tau) := n_\tau B^- e_{id} n_\tau^{-1} e_\tau$ of e_τ in X_w (here, n_τ is a lift in $N(T)$ for the permutation $(a_1 \ldots a_d b_1 \ldots b_{n-d})$, $\{b_1, \ldots, b_{n-d}\}$ being the complement of $\{a_1 \ldots a_d\}$ in $\{1, \ldots, n\}$ arranged in ascending order). We have the following.

Theorem 6.5.7. *Let X_w be a Schubert variety, and $e_\tau \in X_w$.*

1. *The rank of $J_{e_\tau} = \#\{(i,j), 1 \leq j \leq d < i \leq n \mid w \ngeq (\tau(i), \tau(j))\tau\}$ (here, τ is identified with the permutation $(a_1 \ldots a_d b_1 \ldots b_{n-d})$).*
2. $\dim T_{e_\tau}X_w = \#\{(i,j), 1 \leq j \leq d < i \leq n \mid w \geq (\tau(i), \tau(j))\tau\}$.
3. *The Schubert subvariety $X_\tau \subseteq Sing\, X_w$ if and only if*

$$rank J_{e_\tau} < d(n-d) - \dim X_w.$$

Example 6.5.8. The Grassmannian $G_{2,4}$ has one singular Schubert variety: $X_{(2,4)}$. We can see that e_{id} is a singular point of $X_{(2,4)}$ because $X_{(2,4)}$ has dimension 3 and codimension 1, however, $\text{rank}J_{e_{id}} = 0$. On the other hand, if we examine the point $e_{(1,3)}$, $\text{rank}J_{e_{(1,3)}} = 1$ because the action of the reflection $(1,4)$ on $\tau = (1,3)$ results in $(3,4)$, and $(2,4) \not\geq (3,4)$. Therefore, $X_{(1,3)}$ is not in the singular locus of $X_{(2,4)}$. Since every Schubert subvariety of $X_{(2,4)}$ besides $X_{(1,2)}$ contains $X_{(1,3)}$, we have that the singular locus of $X_{(2,4)}$ consists of the point $e_{(1,2)} = X_{(1,2)}$.

In [57, Theorem 5.3] and [87, §5.5], a convenient way of finding $\text{Sing}X_w$ is given, which we shall describe now. For $w = (i_1, \ldots, i_d) \in I_{d,n}$ define a partition $\lambda = (\lambda_1, \ldots, \lambda_d)$ where $\lambda_j = i_{d-j+1} - (d-j+1)$. Thus we have $n-d \geq \lambda_1 \geq \ldots \geq \lambda_d$. We write

$$\lambda = (p_1^{q_1}, \ldots, p_r^{q_r}) = (\underbrace{p_1, \ldots, p_1}_{q_1}, \underbrace{p_2, \ldots, p_2}_{q_2}, \ldots, \underbrace{p_r, \ldots, p_r}_{q_r}).$$

Theorem 6.5.9 (Theorem 5.3 of [57]). *Under the notation above, Sing X_w has $r-1$ components $X_{\alpha_1}, \ldots, X_{\alpha_{r-1}}$, where*

$$\alpha_j = \left(p_1^{q_1}, \ldots, p_{j-1}^{q_{j-1}}, p_j^{q_j-1}, (p_{j+1}-1)^{q_{j+1}+1}, p_{j+2}^{q_{j+2}}, \ldots, p_r^{q_r}\right).$$

If λ is viewed as a Young diagram, then $p_i^{q_i}$ can be viewed as a rectangle. Thus α_j is the Young diagram obtained from λ after deleting a suitable "hook."

Example 6.5.10. In the Grassmannian $G_{3,6}$, let $w = (2,4,6)$, and thus $\lambda = (3,2,1)$. As a Young diagram:

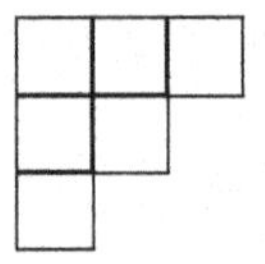

Thus Sing X_w has two components corresponding to $\alpha_1 = (1,1,1)$ and $\alpha_2 = (3,0,0)$:

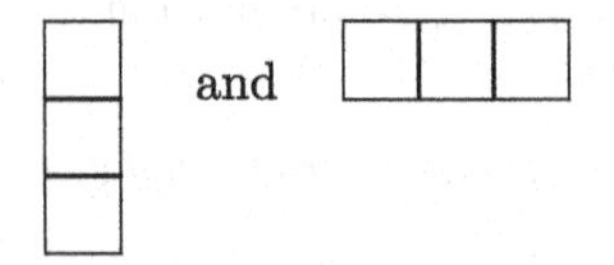

Transforming these partitions back to elements of $I_{3,6}$, we have

$$\text{Sing}\, X_w = X_{(2,3,4)} \cup X_{(1,2,6)}.$$

Remark 6.5.11. We return to geometric properties of Schubert varieties in sections 7.4 and 7.5, after we have introduced some notions concerning distributive lattices.

Chapter 7
Flat Degenerations

We defined flat degenerations in §4.4. In this chapter, we give two examples of flat degenerations of the cone over a Schubert variety. We then explore two additional geometric properties of Schubert varieties: the degree and the property of being Gorenstein; we give a combinatorial characterization for Gorenstein Schubert varieties. We also describe a Gröbner basis for the defining ideal of the Grassmannian variety $G_{d,n}$.

7.1 Gröbner basis

We define a total order on monomials in Plücker coordinates. Let $S=K[p_\tau,\ \tau \in I_{d,n}]$, the polynomial algebra; let I be the defining ideal of the Grassmannian (I is the ideal generated by all quadratic Plücker relations), and let $I(\tau)$ be the defining ideal of X_τ, for the Plücker embedding (the same notation as used in §5.4).

We begin with our standard partial order $\leq$ on $I_{d,n}$. We extend this partial order to a total (dictionary) order, $<^{dct}$: for $\underline{i}, \underline{j} \in I_{d,n}$,

$$\underline{i} <^{dct} \underline{j} \Leftrightarrow i_1 = j_1, \ldots, i_{t-1} = j_{t-1},\ i_t < j_t \text{ for some } 1 \leq t \leq d.$$

We use $<^{dct}$ to define a total order on $\{p_\tau \mid \tau \in I_{d,n}\}$:

$$p_{\tau_1} \succ p_{\tau_2} \Leftrightarrow \tau_1 <^{dct} \tau_2$$

(we have reversed the order for a specific purpose). Using this total order $\succ$ on $\{p_\tau \mid \tau \in I_{d,n}\}$, we use a homogeneous version of the lexicographic order from Definition 4.1.2 to order all monomials in S; namely, given $f = p_{\tau_1} \cdots p_{\tau_r}$ and

V. Lakshmibai, J. Brown, *The Grassmannian Variety*,
Developments in Mathematics 42, DOI 10.1007/978-1-4939-3082-1_7

$g = p_{\mu_1} \cdots p_{\mu_s}$, where $p_{\tau_1} \preceq \ldots \preceq p_{\tau_r}$ and $p_{\mu_1} \preceq \ldots \preceq p_{\mu_s}$, then $f \succ g$ if and only if either $r > s$ or there exists $l < n$ such that $p_{\tau_1} = p_{\mu_1}, \ldots, p_{\tau_l} = p_{\mu_l}$ and $p_{\tau_{l+1}} \succ p_{\mu_{l+1}}$.

Example 7.1.1. Using the terms from Example 5.4.2,

$$p_{(2,3,4)}p_{(1,4,5)} \succ p_{(2,4,5)}p_{(1,3,4)} \succ p_{(3,4,5)}p_{(1,2,4)}.$$

Remark 7.1.2. Suppose $\underline{p}$ is a nonstandard monomial in S, and I as above, the defining ideal of $G_{d,n}$. From Lemma 5.4.4, we know that $\underline{p}$ can be written as a linear combination of standard monomials in S/I, and by Lemma 5.4.6, this linear combination must be unique. Suppose

$$\underline{p} = \sum_{i=1}^{s} a_i \underline{m_i} \pmod{I}$$

such that $a_i \in K$ and $\underline{m_i}$ is a standard monomial for each $1 \leq i \leq s$. Then we will denote $f_{\underline{p}}$ to be the element of S

$$f_{\underline{p}} = \underline{p} - \sum_{i=1}^{s} a_i \underline{m_i}.$$

Note that from Proposition 5.4.12, we have $\underline{p} \succ \underline{m_i}$ for all $1 \leq i \leq s$. Therefore

$$\text{in}_{\succ}(f_{\underline{p}}) = \underline{p}.$$

If $p_\tau p_\mu$ is a nonstandard monomial, let $f_{\tau,\mu}$ denote $f_{p_\tau p_\mu}$ as defined in the previous remark.

Theorem 7.1.3. *The set $\{p_\tau p_\mu \mid p_\tau p_\mu$ nonstandard$\}$ generates $\text{in}_{\succ}(I)$, where I is the ideal generated by the (quadratic) Plücker relations.*

Proof. To begin, we claim that for any $g \in I$, $\text{in}_{\succ}(g)$ is a nonstandard monomial. If possible, let us assume that there exists a $g \in I$ such that $\text{in}_{\succ}(g) = \underline{m_0}$, a standard monomial, and

$$g = a_o \underline{m_0} + \sum_{i=1}^{r} a_i \underline{m_i} + \sum_{j=1}^{t} b_j \underline{n_j}$$

where $\underline{m_i}$ is a standard monomial and $\underline{n_j}$ is a nonstandard monomial, $a_0 \neq 0$. Thus $\underline{m_0} \succ \underline{n_j}$ for all $1 \leq j \leq t$. Now consider $h = g - \sum_{j=1}^{t} b_j f_{\underline{n_j}}$, where $f_{\underline{n_j}}$ is as defined in Remark 7.1.2. Notice that $h \in I$; in addition, h is a linear combination of standard monomials. Since $\underline{m_0} \succ \underline{n_j} = \text{in}_{\succ}(f_{\underline{n_j}})$, we have $\text{in}_{\succ}(h) = \underline{m_0}$, and thus $h \neq 0$. But then h, a (nonzero) sum of standard monomials, is equal to zero in S/I, contradicting Lemma 5.4.6. Therefore, our assumption is wrong and it follows that $\text{in}_{\succ}(g)$ is nonstandard for all $g \in I$.

Next, we show that any nonstandard monomial is in the ideal generated by nonstandard monomials of degree 2. Let $p_{\tau_1} \cdots p_{\tau_r}$ be a nonstandard monomial. Then there exists some $l < r$ such that $p_{\tau_1} \geq p_{\tau_2} \geq \ldots \geq p_{\tau_{l-1}} \not\geq p_{\tau_l}$. Thus, this monomial is an element of the ideal generated by $p_{\tau_{l-1}} p_{\tau_l} (= \text{in}_{\succ}(f_{\tau_{l-1},\tau_l}))$. The required result now follows. □

7.2 Toric Degenerations

In this section, we exhibit a flat degeneration of the cone over a Schubert variety to an affine toric variety. Here, we give some basic definitions related to toric varieties; we recommend [22, 37] for a more detailed treatment. An *n-dimensional algebraic torus* is $(K^*)^n$, for K a field. The group of invertible $n \times n$ diagonal matrices is an example of an n-dimensional torus.

Definition 7.2.1. An *equivariant affine embedding* of a torus T (or also an *affine toroidal embedding*) is an affine variety X containing T as a dense open subset, together with a T-action $T \times X \to X$ extending the action $T \times T \to T$ given by multiplication. In addition, if X is normal, then X is called an *affine toric variety*.

We will show that an affine variety defined by a *binomial prime ideal* (i.e., an ideal with generators consisting of polynomials with consisting of two monomials) is an affine toroidal embedding. For $N \geq 1$, let X be an affine variety in $\mathbb{A}^N = \text{Spec}K[x_1, \ldots, x_N]$, such that X is not contained in any coordinate hyperplanes given by $\{x_i = 0\}$. In addition, let X be irreducible, and let $\mathcal{I}(X)$ be a prime ideal generated by the l binomials

$$x_1^{a_{i1}} \ldots x_N^{a_{iN}} - \lambda_i x_1^{b_{i1}} \ldots x_N^{b_{iN}} \quad 1 \leq i \leq l. \tag{\dagger}$$

Let T_N be the maximal torus in $\mathbb{A}^N$, $T_N = \{(t_1, \ldots, t_N) \mid t_i \neq 0, \ \forall\, i\}$. We have the natural action of T_N on $\mathbb{A}^N$: $(t_1, \ldots, t_N) \cdot (c_1, \ldots, c_N) = (t_1 c_1, \ldots, t_N c_N)$ (and hence the affine space is a toric variety). Consider $X(T_N) = \text{Hom}(T_N, K^*)$, the character group of T_N. Define $\varepsilon_i \in X(T_N)$ such that $\varepsilon_i(t_1, \ldots, t_N) = t_i$, $1 \leq i \leq N$; and for $1 \leq i \leq l$, define $\phi_i \in X(T_N)$ as such:

$$\phi_i = \sum_{t=1}^{N} (a_{it} - b_{it})\, \varepsilon_t$$

(with $\{a_{it}, b_{it}\}$ as in (†)).

Now set $T = \bigcap_{i=1}^{l} \ker\phi_i \subseteq T_N$ and $X^0 = \{(x_1, \ldots, x_N) \in X \mid x_i \neq 0 \, \forall\, i\} (= X \cap T_N)$. We have the following proposition.

Proposition 7.2.2.

1. *We have a canonical action of T on X.*
2. *The subset X^0 is T-stable, and the action of T on X^0 is simple and transitive.*
3. *T is a subtorus of T_N, and X is an affine toroidal embedding.*

Proof. We have the (obvious) action of T on $\mathbb{A}^N$; we shall now show that for $t \in T$ and $x \in X$, $t \cdot x \in X$. Let $t = (t_1, \ldots, t_N)$ and $x = (x_1, \ldots, x_N)$, and suppose $(t_1x_1, \ldots, t_Nx_N) = (y_1, \ldots, y_N)$. We have that

$$x_1^{a_{i1}} \ldots x_N^{a_{iN}} = \lambda_i x_1^{b_{i1}} \ldots x_N^{b_{iN}} \quad 1 \leq i \leq l,$$

and thus

$$t_1^{a_{i1}} \ldots t_N^{a_{iN}} x_1^{a_{i1}} \ldots x_N^{a_{iN}} = \lambda_i t_1^{b_{i1}} \ldots t_N^{b_{iN}} x_1^{b_{i1}} \ldots x_N^{b_{iN}}$$
$$y_1^{a_{i1}} \ldots y_N^{a_{iN}} = \lambda_i y_1^{b_{i1}} \ldots y_N^{b_{iN}} \quad 1 \leq i \leq l.$$

Therefore $(y_1, \ldots, y_N) \in X$ and (1) follows.

Clearly, (1) implies that $t \cdot x \in X^0$ for $t \in T$ and $x \in X^0$, since $X^0 = X \cap T_N$. Now, given $x \in X^0$, the isotropy subgroup of T_N at x is just $\{e\}$, e being the identity element $(1, \ldots, 1) \in T_N$, and hence the isotropy subgroup of T at x is also just $\{e\}$. Hence the action is simple. To see that the action is transitive, let $x, x' \in X^0$; define t such that $t_i = x_i/x'_i$. Then $t \cdot x' = x$. This completes the proof of (2).

To prove (3), we first observe that for $x \in X^0$, the orbit map $t \mapsto t \cdot x$ is an isomorphism of T onto X^0 (in view of (2)). Define $X_i = \{(x_1, \ldots, x_N) \in X \mid x_i \neq 0\}$. In view of the hypothesis that X is not contained in any of the coordinate hyperplanes, we have that X_i is nonempty for $1 \leq i \leq N$. Because X is irreducible, X_i is open and dense; thus $X^0 = \bigcap X_i$ is open, dense, and irreducible in X. Hence, T is irreducible (and hence connected). Thus T is a subtorus of T_N and (3) follows. □

The results above summarize into the following:

Theorem 7.2.3. *An affine variety X such that $\mathcal{I}(X)$ is prime and generated by binomials is an affine toroidal embedding.*

Remark 7.2.4. In fact, in [86, Lemma 4.1] it is shown that $\mathcal{I}(X)$ being prime and generated by binomials is necessary for an affine variety X to be a toroidal embedding. Thus we have a one-to-one correspondence:

$$\{\text{affine toroidal embeddings}\} \leftrightarrow \{\text{binomial prime ideals}\}$$

One example of a toric variety that is relevant to our discussion of flat deformations of toric varieties is a "Hibi toric variety," (which we define below), based upon a given distributive lattice.

Definition 7.2.5. A *lattice* is a partially ordered set H such that, for every pair of elements $x, y \in H$, there exist elements $x \vee y$ and $x \wedge y$, called the *join*, respectively the *meet*, of x and y, defined by:

$$x \vee y \geq x,\ x \vee y \geq y, \text{ and if } z \geq x \text{ and } z \geq y, \text{ then } z \geq x \vee y,$$
$$x \wedge y \leq x,\ x \wedge y \leq y, \text{ and if } z \leq x \text{ and } z \leq y, \text{ then } z \leq x \wedge y.$$

The meet and join operations are clearly commutative and associative. A pair of two noncomparable elements of a lattice is called a *skew pair*. Note that if x and y are comparable, say $x \leq y$, then by the definition above $x \vee y = y$ and $x \wedge y = x$.

Definition 7.2.6. A lattice is called *distributive* if the following identities hold:

$$x \wedge (y \vee z) = (x \wedge y) \vee (x \wedge z)$$
$$x \vee (y \wedge z) = (x \vee y) \wedge (x \vee z).$$

Example 7.2.7. For any natural numbers $d < n$, the partially ordered set $I_{d,n}$ is a distributive lattice. In addition, for any $w \in I_{d,n}$; the subset $\{\tau \in I_{d,n} \mid \tau \leq w\}$ is also a distributive lattice. Let us look specifically at $I_{2,4}$:

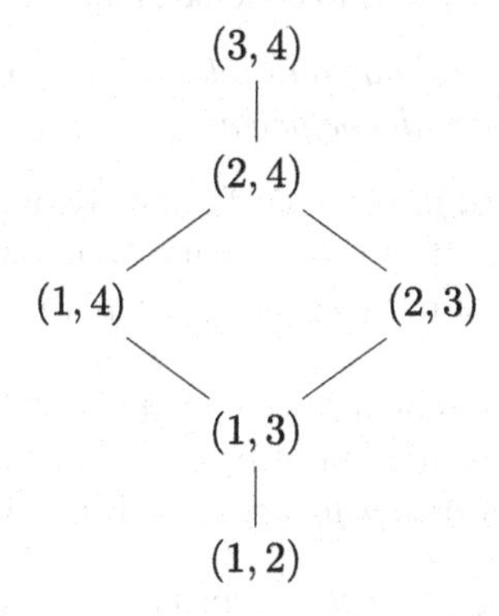

The only skew pair in $I_{2,4}$ is $(1,4)$ and $(2,3)$. One can see by the diagram that $(1,4) \vee (2,3) = (2,4)$ and $(1,4) \wedge (2,3) = (1,3)$. For a general $I_{d,n}$, $\underline{i} \vee \underline{j} = (k_1, \dots, k_d)$ and $\underline{i} \wedge \underline{j} = (l_1, \dots, l_d)$, where $k_t = \max\{i_t, j_t\}$ and $l_t = \min\{i_t, j_t\}$.

Let H be a distributive lattice, and let S be the polynomial algebra $S = K[x_\tau,\ \tau \in H]$. Let $I(H)$ be the ideal of S generated by

$$\{x_\alpha x_\beta - x_{\alpha \vee \beta} x_{\alpha \wedge \beta} \mid \alpha, \beta \in H\}.$$

Note that $I(H)$ is a binomial ideal. We cite the following theorem without proof [30, p. 100].

Theorem 7.2.8. *With notation as above, $S/I(H)$ is a normal, integral domain.*

Since $S/I(H)$ is an integral domain, $I(H)$ is a prime binomial ideal, and thus $S/I(H)$ is an affine toroidal embedding. Since $S/I(H)$ is normal, we have that $\operatorname{Spec} S/I(H)$ is a toric variety. We call such a toric variety a *Hibi toric variety*.

Given a Schubert variety X_w, we shall construct a flat family, with $\widehat{X_w}$ (the cone over $X(w)$) as the generic fiber and a Hibi toric variety as the special fiber. We first gather some facts.

Remark 7.2.9. Recall from Proposition 5.4.12 that given a straightening relation for a skew pair τ, ϕ:

$$p_\tau p_\phi = \sum_{(\alpha,\beta)} c_{\alpha,\beta} p_\alpha p_\beta, \quad c_{\alpha,\beta} \in K^*, \alpha > \beta, \qquad (*)$$

we have that $\alpha > \tau, \phi$, for any α appearing on the right hand side. Here we note that we also have $\beta < \tau, \phi$ for any β appearing on the right hand side. To see this, let $w_0 \in S_n$ be the permutation $(n, n-1, \dots, 1)$. Recall from §6.3 that we have an action of S_n on elements of $I_{d,n}$. We can see that if we replace each $x \in I_{d,n}$ appearing in $(*)$ by w_0x, we still have a straightening relation, in fact the action of w_0 reverses the order in $I_{d,n}$; i.e., $\alpha > \beta$ implies $w_0\alpha < w_0\beta$. Therefore we can apply the proof of Proposition 5.4.12 to $(p_{w_0\tau}, p_{w_0\phi})$ to conclude that $\beta < \tau, \phi$.

Lemma 7.2.10. *Given the straightening relation in* $(*)$, *the monomial* $p_{\tau\vee\phi}p_{\tau\wedge\phi}$ *occurs on the right hand side with coefficient 1.*

Proof. Denote $\lambda = \tau \vee \phi$ and $\mu = \tau \wedge \phi$. Then by Remark 7.2.9 and the definition of join and meet, we have that for any $\alpha > \beta$ that occur on the right hand side of $(*)$, $\alpha \geq \lambda$ and $\beta \leq \mu$; further, if $\alpha = \lambda$, then $\beta = \mu$ (by weight considerations, cf. §5.2.3).

We first show that $p_\lambda p_\mu$ does in fact occur in $(*)$. If we restrict $(*)$ to the Schubert variety X_λ, we see that the restriction of $p_\tau p_\phi \neq 0$, and by the discussion above, restrictions of all terms other than $p_\lambda p_\mu$ are zero. Thus we have on X_λ

$$p_\tau p_\phi = c p_\lambda p_\mu,$$

where $c \neq 0$ (since, $K[X_\lambda]$ is an integral domain).

Since the standard monomial basis is characteristic free, we conclude that $c = \pm 1$; we shall now show that $c = 1$. Let $\tau = (i_1, \dots, i_d)$ and $\phi = (j_1, \dots, j_d)$, and thus $\lambda = (k_1, \dots, k_d)$ and $\mu = (l_1, \dots, l_d)$ as defined in Example 7.2.7. Now consider a pair $(\alpha > \beta)$ occurring in $(*)$, $\alpha \neq \lambda$. Let $\alpha = (a_1, \dots, a_d), \beta = (b_1, \dots, b_d)$; since $\alpha > \lambda$, let s be the smallest number such that $a_s > k_s$. Thus $a_t = k_t$ for all $t \leq s-1$. Using the fact that $\tau \dot{\cup} \phi = \alpha \dot{\cup} \beta$ (by weight considerations), we have

$$a_p = k_p, b_p = l_p, p < s, \ k_s \notin \{a_1, \dots, a_d\}, \ k_s, l_s \in \{b_1, \dots, b_d\}$$

Now, given a generic $n \times d$-matrix $A = (x_{ij})$, we have $p_\tau(A) = \det A_{\underline{i}}$, the determinant of the d-minor of A with rows given by $(i_1, \dots, i_d)$ (cf. §5.2.1). Let us now compare the coefficient of the monomial $\underline{m} := x_{i_1 1} \dots x_{i_d d} x_{j_1 1} \dots x_{j_d d}$ on both sides of $(*)$. Clearly, $\underline{m}$ being the product of respective diagonal terms of $p_\tau(A)$ and $p_\phi(A)$, occurs with coefficient 1 on the left hand side of $(*)$. Also note that the product $x_{l_s s} x_{k_s s}$ appears in $\underline{m}$. Now the fact that $l_s,\ k_s \in \beta$ implies that the product $x_{l_s s} x_{k_s s}$ does not appear in $p_\alpha(A) p_\beta(A)$, for all $\alpha > \lambda$. This implies that $\underline{m}$ should occur with coefficient 1 in $p_\lambda p_\mu$ (in fact, in $p_\lambda p_\mu$, $\underline{m}$ is realized as the product $\underline{m}_1 \underline{m}_2$, where $\underline{m}_1$ (resp. $\underline{m}_2$) is the product of the diagonal terms of $p_\lambda(A)$ (resp.$p_\mu(A)$). From this it follows that the coefficient $c = 1$. □

For the discussion that follows, let $w \in I_{d,n}$, and let $H = \{\tau \in I_{d,n} \mid \tau \leq w\}$. Let $S_H = K[x_\tau,\ \tau \in H]/I(H)$ and let $R(w)$ be the homogeneous coordinate ring of X_w. We have a surjective map

$$\pi : K[x_\theta,\ \theta \in H] \to R(w)$$

$$x_\theta \mapsto p_\theta.$$

We have $\ker\pi = \langle f_{\tau,\phi},\ (\tau,\phi) \text{ skew pair}\rangle$, where

$$f_{\tau,\phi} = x_\tau x_\phi - \sum c_{\alpha\beta} x_\alpha x_\beta$$

as in $(*)$ from Remark 7.2.9.

Fix a sufficiently large N, and for $\theta = (a_1, \ldots, a_d) \in H$, let

$$N_\theta = \sum_{r=1}^{d} N^{d-r} a_r,$$

$a_1, \ldots, a_d$ being the integers in the N-ary presentation of N_θ. Let $A = K[t]$, $P_A = A[x_\theta,\ \theta \in H]$, and Q be the set of all skew pairs in H. For $(\tau,\phi) \in Q$, define $f_{\tau,\phi,t}$ in P_A as

$$f_{\tau,\phi,t} = x_\tau x_\phi - \sum c_{\alpha\beta} x_\alpha x_\beta t^{N_\alpha + N_\beta - N_\tau - N_\phi}.$$

For the definition above to be valid, we must show that $N_\alpha + N_\beta - N_\tau - N_\phi \geq 0$. Let us use the notation from the proof of Lemma 7.2.10 concerning $\phi, \tau, \lambda, \mu, \alpha$, and β. We note that for $1 \leq t \leq d$, we have $\{k_t, l_t\} = \{i_t, j_t\}$, and thus $k_t + l_t = i_t + j_t$; therefore $N_\lambda + N_\mu = N_\phi + N_\tau$. On the other hand, for $\alpha > \lambda$, let s be such that $a_1 = k_1, \ldots, a_{s-1} = k_{s-1}$ but $a_s > k_s$. As shown in the proof of Lemma 7.2.10, we have $b_1 = l_1, \ldots, b_s = l_s$. Thus $a_s + b_s > k_s + l_s$, implying $N_\alpha + N_\beta > N_\lambda + N_\mu = N_\tau + N_\phi$.

Now let $\mathcal{I}$ be the ideal in P_A generated by $\{f_{\tau,\phi,t},\ (\tau,\phi) \in Q\}$, and $\mathcal{R} = P_A/\mathcal{I}$.

Lemma 7.2.11. *With notation as above,*

1. *$\mathcal{R}$ is $K[t]$-free.*
2. *$\mathcal{R} \otimes_{K[t]} K[t, t^{-1}] \cong R(w)[t, t^{-1}]$.*
3. *$\mathcal{R} \otimes_{K[t]} K[t]/\langle t\rangle \cong S_H$.*

Proof. We first note that $\mathcal{R} \otimes_{K[t]} K[t]/\langle t\rangle = P_A/(\mathcal{I} + \langle t\rangle)$, which is clearly isomorphic to S_H and (3) follows.

Let $B = K[t, t^{-1}]$, $P_B = B[x_\theta, \theta \in H]$. Let $\bar{I}$ be the ideal in P_B generated by $\{f_{\tau,\phi}, (\tau, \phi) \in Q\}$ and $\bar{\mathcal{I}}$ be the ideal in P_B generated by $\{f_{\tau,\phi,t}, (\tau, \phi) \in Q\}$. We have

$$P_B/\bar{I} \cong R(w)[t, t^{-1}], \qquad (\dagger)$$

$$P_B/\bar{\mathcal{I}} \cong \mathcal{R} \otimes_{K[t]} K[t, t^{-1}]. \qquad (\dagger\dagger)$$

The automorphism $P_B \simeq P_B$, $x_\alpha \mapsto t^{N_\alpha} x_\alpha$ induces an isomorphism

$$P_B/\bar{I} \cong P_B/\bar{\mathcal{I}}. \qquad (\dagger\dagger\dagger)$$

Assertion (2) follows from (†), (††), and (†††).

It remains to prove (1). We use the following notation:

$$X_\alpha = \bar{x}_\alpha \in \mathcal{R},$$

$$P_\alpha = t^{N_\alpha} X_\alpha,$$

$$\mathcal{M} = \{P_{\alpha_1} \cdots P_{\alpha_r}, \alpha_1 \geq \ldots \geq \alpha_r, r \in \mathbb{Z}_+\}$$

$$\mathcal{N} = \{X_{\alpha_1} \cdots X_{\alpha_r}, \alpha_1 \geq \ldots \geq \alpha_r, r \in \mathbb{Z}_+\}$$

Claim. $\mathcal{N}$ is a $K[t]$ basis for $\mathcal{R}$.

Since standard monomials form a K-basis for $R(w)$, we have, by base change, standard monomials form a $K[t, t^{-1}]$-basis for $R(w)[t, t^{-1}]$. Now under the isomorphism

$$\phi : R(w)[t, t^{-1}] \to \mathcal{R} \otimes_{K[t]} K[t, t^{-1}] (= \mathcal{R}[t^{-1}])$$

$$p_\alpha \mapsto t^{N_\alpha} X_\alpha = P_\alpha$$

the standard monomial $p_{\alpha_1} \ldots p_{\alpha_r}$ gets mapped to $P_{\alpha_1} \ldots P_{\alpha_r}$. Hence $\mathcal{M}$ is a $K[t, t^{-1}]$-basis for $\mathcal{R}[t^{-1}]$. In particular, $\mathcal{M}$ is linearly independent over $K[t, t^{-1}]$, and hence $\mathcal{N}$ is linearly independent over $K[t]$.

Next, we show generation; i.e., that monomials in X_α of degree r are spanned by degree r elements of $\mathcal{N}$. Let $F = X_{\tau_1} \ldots X_{\tau_r} \in \mathcal{R}$ such that there exists i such that $\tau_i \not\geq \tau_{i+1}$; denote $\tau_i = \tau$, $\tau_{i+1} = \phi$. Since $\phi(p_\theta) = P_\theta$, we have the relation

$$P_\tau P_\phi = \sum c_{\alpha\beta} P_\alpha P_\beta$$

$$\text{i.e., } X_\tau X_\phi = \sum c_{\alpha\beta} t^{N_\alpha + N_\beta - N_\tau - N_\phi} X_\alpha X_\beta.$$

We substitute for $X_{\tau_i} X_{\tau_{i+1}}$ to obtain $F = \sum a_i F_i$. Define

$$\mathrm{wt}(F) = l(\tau_1) N^{r-1} + \ldots + l(\tau_{r-1}) N + l(\tau_r),$$

where $l(\tau) = \dim X_\tau$. We have $\mathrm{wt}(F_i) > \mathrm{wt}(F)$. Therefore, by decreasing induction on $\mathrm{wt}(F)$, we have that each F_i is a $K[t]$-linear combination of elements of $\mathcal{N}$. Thus the claim follows.

The proof of the claim completes the proof of (1). □

The lemma above leads to the following result.

Theorem 7.2.12. *Spec* $\mathcal{R}$ *is a flat family over Spec* $K[t]$ *whose generic fiber is Spec* $R(w)$ *and special fiber is the Hibi toric variety* S_H.

We note that Spec $R(w)$ is the cone over the Schubert variety X_w. Thus, for $w \in I_{d,n}$, we have a flat family in which $\widehat{X_w}$ is the generic fiber, and the Hibi toric variety S_H is the special fiber, where H is the distributive lattice of Schubert subvarieties of X_w.

Example 7.2.13. Consider the distributive lattice $I_{2,5}$:

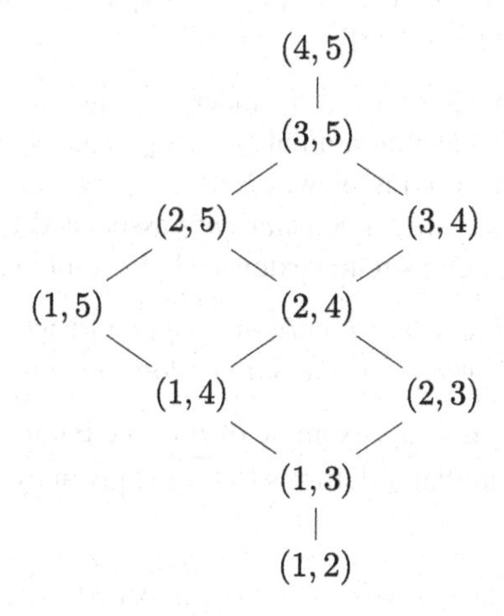

The Hibi toric variety that flatly deforms to $\widehat{G_{2,5}}$ is given by Spec$K[x_\tau,\ \tau \in I_{2,5}]/I$, where I is generated by the binomials

$$x_{(1,4)}x_{(2,3)} - x_{(1,3)}x_{(2,4)},\ x_{(1,5)}x_{(2,3)} - x_{(1,3)}x_{(2,5)},$$
$$x_{(1,5)}x_{(2,4)} - x_{(1,4)}x_{(2,5)},\ x_{(1,5)}x_{(3,4)} - x_{(1,4)}x_{(3,5)},$$
$$x_{(2,5)}x_{(3,4)} - x_{(2,4)}x_{(3,5)}.$$

7.3 Monomial Scheme Degenerations

In this subsection, we show that the cone over a Schubert variety degenerates to a monomial scheme. We change one piece of notation from the previous section: for $\theta \in H$, let $N_\theta = 3^{\dim X_\theta}$. For $(\tau, \phi) \in Q$ (where Q is the set of skew pairs in H), define $f_{\tau,\phi,t}$ in P_A as

$$f_{\tau,\phi,t} = x_\tau x_\phi - \sum c_{\alpha\beta} x_\alpha x_\beta t^{N_\alpha+N_\beta-N_\tau-N_\phi}.$$

We again check that $N_\alpha + N_\beta - N_\tau - N_\phi \geq 0$. We have $\dim X_\tau, \dim X_\phi \leq \dim X_\alpha - 1$. Therefore $N_\tau \leq 3^{\dim X_\alpha - 1}$. Thus

$$N_\tau + N_\phi \leq 2 \cdot 3^{\dim X_\alpha - 1} = \frac{2}{3} N_\alpha < N_\alpha.$$

Therefore $N_\alpha + N_\beta - N_\tau - N_\phi > 0$.

Define $\mathcal{R}$ as above, namely $\mathcal{R} = P_A/\mathcal{I}$, where $\mathcal{I}$ is generated by the set $\{f_{\tau,\phi,t}, (\tau,\phi) \in Q\}$. We have, as above, the following lemma.

Lemma 7.3.1.

1. *$\mathcal{R}$ is $K[t]$-free.*
2. *$\mathcal{R} \otimes_{K[t]} K[t,t^{-1}] \cong R(w)[t,t^{-1}]$.*
3. *$\mathcal{R} \otimes_{K[t]} K[t]/\langle t\rangle \cong S$, where S is $K[x_\theta, \theta \in H]/I$, I being the monomial ideal generated by $\{x_\tau x_\phi, (\tau,\phi) \in Q\}$.*

Thus denoting the image of x_α in S (under the canonical surjection $K[x_\alpha, \alpha \in H] \to S$) by $\bar{x}_\alpha$, we have that a monomial $\bar{x}_{\alpha_1} \ldots \bar{x}_{\alpha_r}$ is nonzero if and only if it does not contain a $\bar{x}_\tau \bar{x}_\phi$ where (τ,ϕ) is skew.

This S is the so-called *Stanley-Reisner ring* associated to the partially ordered set H. (These rings have been studied extensively in combinatorics.)

Theorem 7.3.2. *Spec $\mathcal{R}$ is a flat family over Spec $K[t]$ whose generic fiber is $\widehat{X_w}$ and special fiber Spec S, where S is the Stanley-Reisner ring described above.*

Example 7.3.3. Returning to the example of $I_{2,5}$ (see Example 7.2.13), we see that the monomial scheme that flatly deforms to $\widehat{G_{2,5}}$ is given by Spec$K[x_\tau, \tau \in I_{2,5}]/I$, where

$$I = \left\langle x_{(1,4)}x_{(2,3)}, x_{(1,5)}x_{(2,3)}, x_{(1,5)}x_{(2,4)}, x_{(1,5)}x_{(3,4)}, x_{(2,5)}x_{(3,4)} \right\rangle.$$

7.4 Application to the Degree of X_w

In this section, we show that the monomial scheme degenerations of the Grassmannian variety and Schubert varieties discussed above give us a convenient way to calculate the degree of these varieties. The degree of a projective variety can be thought of as the number of points in which it intersects a general plane of complimentary dimension; but we recall its precise definition here.

Let B be a $\mathbb{Z}_+$-graded, finitely generated K-algebra, $B = \oplus B_m$, with $B_0 = K$. Let $\phi_B(m)$ denote the *Hilbert function*, defined for $m \geq 0$ as:

$$\phi_B(m) = \dim_K B_m.$$

Theorem 7.4.1 (Theorem 7.5, Chap. I of [28]). *The Hilbert function $\phi_B(m)$ agrees for large m with a unique polynomial function with rational coefficients.*

Definition 7.4.2. The polynomial of the previous theorem is called the *Hilbert polynomial* of B, denoted $P_B(x)$.

Remark 7.4.3. We summarize the following results about the Hilbert polynomial, see [28, §I.7] for proof.

1. $P_B(x) \in \mathbb{Q}[x]$.
2. Deg $P_B(x) = \text{dimProj}B = s$, say.
3. The leading coefficient of $P_B(x)$ is of the form $\frac{e_B}{s!}$, where e_B is a positive integer.

Definition 7.4.4. The number e_B from the previous remark is called the *degree* of the graded ring B, or also the degree of Proj B.

Remark 7.4.5. In [28, §III Theorem 9.9], it is shown that in a flat family of projective varieties, the Hilbert polynomial is the same for all the members of the family, and hence so is the degree. We now show a straightforward way to compute the degree of the monomial scheme degenerations introduced in §7.3, which will give us the degree of any Schubert variety (for the Plücker embedding).

For the remark and theorem that follow, let $\mathcal{L}$ be a distributive lattice, let $R = K[x_\tau,\ \tau \in \mathcal{L}]$, and let $I(\mathcal{L})$ be the ideal of R generated by the set of monomials $\{x_\alpha x_\beta \mid (\alpha, \beta) \text{ a skew pair in } \mathcal{L}\}$. Then $S = R/I(\mathcal{L})$ is the Stanley-Reisner ring mentioned in §7.3.

Remark 7.4.6. The Krull dimension of S is the maximal number of elements in a *chain* of $\mathcal{L}$ (by chain, we mean a totally ordered subset of $\mathcal{L}$). To see this, note that for any skew pair (α, β) in $\mathcal{L}$, any prime ideal of S must contain either $\bar{x}_\alpha$ or $\bar{x}_\beta$ (where $\bar{x}_\alpha \in S$ represents the image of $x_\alpha \in R$). Thus, if P_0 is a minimal prime ideal of S, and $\{\bar{x}_{\alpha_1}, \ldots, \bar{x}_{\alpha_n}\}$ are the degree one terms of $S \setminus P_0$, then $\{\alpha_1, \ldots, \alpha_n\}$ form a maximal, totally ordered set in $\mathcal{L}$.

Theorem 7.4.7. *With notation as above, the degree of S is equal to the number of maximal chains in $\mathcal{L}$.*

Proof. Let $J = \{j_1, \ldots, j_s\}$ be a subset of $\mathcal{L}$ such that $X_{j_1} \cdots X_{j_s} \notin I(\mathcal{L})$. Note that J is thus a chain in $\mathcal{L}$. We have

$$S = K \oplus \bigoplus_{J=\{j_1,\ldots,j_s\}} (X_{j_1} \cdots X_{j_s})\, K[X_{j_1}, \ldots, X_{j_s}],$$

where J runs over all chains of any length in $\mathcal{L}$. Therefore, we have

$$\begin{aligned} \phi_S(m) &= \dim S_m \\ &= \sum_{J=\{j_1,\ldots,j_s\}} \binom{s + (m-s) - 1}{m-s} \\ &= \sum_{J=\{j_1,\ldots,j_s\}} \binom{m-1}{m-s}. \end{aligned}$$

Note that for m sufficiently large, the leading term (of the polynomial with respect to m) appears in the summation above only for J of maximal cardinality s. The computation above can also be written as

$$\sum_{J=\{j_1,\ldots,j_s\}} \binom{m-1}{s-1}.$$

From this, we have that the leading coefficient is of the form $\frac{e}{(s-1)!}$, where $\dim \mathrm{Proj} S = s-1$ (s being the number of elements in a maximal chain of $\mathcal{L}$), and e is the number of maximal chains J. The result follows. □

Corollary 7.4.8. *The degree of $G_{d,n}$ is the number of maximal chains in $I_{d,n}$. For $w \in I_{d,n}$, the degree of X_w is the number of maximal chains in H_w, where*

$$H_w = \{\alpha \in I_{d,n} \mid \alpha \leq w\}.$$

Proof. The corollary follows from Theorem 7.4.7 and Remark 7.4.5. □

Example 7.4.9. From the figure of $I_{2,4}$ in Example 7.2.7, we can easily see that the degree of $G_{2,4}$ is 2, as is the degree of the Schubert variety $X_{(2,4)}$. The other four Schubert varieties of $G_{2,4}$ have degree 1.

We now derive an explicit formula for the number of maximal chains in $I_{d,n}$, and hence the degree of $G_{d,n}$. Notice that the number of chains in $I_{d,n}$ from $(1,2,\ldots,d)$ to $(n-d+1,\ldots,n)$ is the same as the number of chains from $(0,0,\ldots,0)$ to $(n-d,n-d,\ldots,n-d)$ such that for any $(i_1,\ldots,i_d)$ in the chain, we have $i_1 \geq i_2 \geq \ldots \geq i_d \geq 0$. Now, set

$$\mu = (\mu_1,\mu_2,\ldots,\mu_d) = (n-d,n-d,\ldots,n-d). \qquad (*)$$

For any partition λ of m (i.e., $\lambda = (\lambda_1,\ldots,\lambda_l)$ such that $\lambda_1 + \cdots + \lambda_l = m$), let f^λ be the *Kostka number* $K_{\lambda,1^m}$, i.e., the number of standard Young tableau of shape λ (we define standard Young tableau in Definition 8.2.3).

Proposition 7.4.10 (cf. Proposition 7.10.3, [85]). *Let $\lambda = (\lambda_1,\ldots,\lambda_l)$ be a partition of m. Then the number f^λ counts the lattice paths $0 = v_0, v_1,\ldots,v_m$ in $\mathbb{R}^l$ from the origin v_0 to $v_m = (\lambda_1,\lambda_2,\ldots,\lambda_l)$, with each step a coordinate vector; and staying within the region (or cone) $x_1 \geq x_2 \geq \ldots \geq x_l \geq 0$. (By "each step a coordinate vector," we mean that if one element of the lattice path is $(x_1,\ldots,x_l)$, then the next element is of the form $(x_1,\ldots,x_{i-1},x_i+1,x_{i+1},\ldots,x_l)$ for some $1 \leq i \leq l$.)*

Thus, for μ as described in $(*)$ above, we have that the number of maximal chains in $I_{d,n}$ is equal to f^μ.

Example 7.4.11. For $n = 4,\ d = 2$, there are only two lattice paths from $(0,0)$ to $(2,2)$ satisfying the description above:

$$\{(0,0),\ (1,0),\ (2,0),\ (2,1),\ (2,2)\},\ \text{and}\ \{(0,0),\ (1,0),\ (1,1),\ (2,1),\ (2,2)\}$$

Corollary 7.21.5 of [85] gives an explicit description of f^λ.

Proposition 7.4.12. *Let $\lambda = (\lambda_1, \ldots, \lambda_l)$ be a partition of m. Then*

$$f^\lambda = \frac{m!}{\prod_{u\in\lambda} h(u)}.$$

The statement above refers to $u \in \lambda$ as a box in the Young diagram of λ, and $h(u)$ being the "hook length" of u. The hook length is defined as the number of boxes to the right, and below, of u, including u once.

Example 7.4.13. Let us take for example $I_{3,6}$. Then $\mu = (3,3,3)$, and the Young diagram of shape μ with hook lengths given in their corresponding boxes is

5	4	3
4	3	2
3	2	1

Therefore

$$f^\mu = \frac{9!}{5 \cdot 4^2 \cdot 3^3 \cdot 2^2 \cdot 1} = 42.$$

Thus $\deg G_{3,6} = 42$.

In fact, in the scenario of $I_{d,n}$, our derived partition μ (given by $(*)$) will always be a rectangle; and we can deduce a formula for f^μ which does not require the Young tableau. The top left box of μ will always have hook length

$$(n-d) + d - 1 = n - 1.$$

Then, the box directly below it, and the box directly to the right of it will have length $n-2$. For any box of μ, the box below and the box to the right will have hook length one less than that of the box with which we started.

Since the posets $I_{d,n}$ and $I_{n-d,n}$ are isomorphic, we may assume that $d \le n-d$. Then we have

$$\prod_{u\in\mu} h(u) = (n-1)(n-2)^2 \cdots (n-d)^d (n-d-1)^d \cdots (d)^d (d-1)^{d-1} \cdots (2)^2(1).$$

Thus we arrive at the following.

Theorem 7.4.14. *The degree of $G_{d,n}$ is equal to*

$$\frac{(d\,(n-d))!}{(n-1)(n-2)^2\cdots(n-d)^d(n-d-1)^d\cdots(d)^d(d-1)^{d-1}\cdots(2)^2(1)}.$$

Equivalently,

$$\deg G_{d,n} = (d\,(n-d))!\prod_{i=0}^{d-1}\frac{i!}{(n-d+i)!}.$$

Remark 7.4.15. We note that for a Grassmannian variety of the form $G_{2,n}$, the degree happens to be one of the ubiquitous Catalan numbers,

$$\deg G_{2,n} = \mathrm{Cat}_{n-2} = \frac{1}{n-1}\binom{2n-4}{n-2}.$$

While we will not give an explicit formula, the methods described above also apply to the degree of a Schubert variety X_w. For $w = (w_1, \ldots, w_d) \in I_{d,n}$, define $\lambda = (w_d - d, w_{d-1} - (d-1), \ldots, w_1 - 1)$. We may use Proposition 7.4.12 to calculate the degree of X_w.

Example 7.4.16. Consider X_w in $G_{3,6}$, where $w = (2, 4, 6)$. Then $\lambda = (3, 2, 1)$. As a Young tableau, with hook lengths given in their corresponding boxes, we have

5	3	1
3	1	
1		

Thus $f^\lambda = \dfrac{6!}{5\cdot 3^2\cdot 1^3} = 16$. Therefore $\deg X_{(2,4,6)} = 16$.

7.5 Gorenstein Schubert Varieties

In this section, we define the Gorenstein property for an algebraic variety; we then give a combinatorial characterization for Gorenstein Schubert varieties. In general, the Gorenstein property is a geometric property between Cohen–Macaulayness and smoothness; i.e., every Gorenstein variety is Cohen–Macaulay, but not every Gorenstein variety is smooth, whereas every smooth variety is Gorenstein. Recall that in §6.1, we saw that every Schubert variety is Cohen–Macaulay; whereas in Theorem 6.4.2, we saw that the only smooth Schubert varieties are Schubert varieties which are isomorphic to a Grassmannian variety.

Let R be a commutative ring, and M, N be R-modules. Recall from Proposition 3.3.7 that the category of R-modules has enough injectives, and thus by Proposition 3.2.11 there exist injective resolutions of M and N. Let $I_\bullet$ be an injective resolution of N:

$$0 \to N \to I_0 \to I_1 \to \cdots .$$

We now apply the functor $\mathrm{Hom}_R(M,-)$ to the injective resolution, obtaining the cochain complex $\mathrm{Hom}_R(M, I_\bullet)$:

$$0 \to \mathrm{Hom}_R(M, I_0) \to \mathrm{Hom}_R(M, I_1) \to \cdots .$$

As in Definition 3.2.12, we form the right derived functor

$$R^i\left(\mathrm{Hom}_R(M,-)\right)(N),$$

and we denote it $\mathrm{Ext}^i_R(M,N)$.

The definition of a Gorenstein ring goes hand-in-hand with the definition of a Cohen–Macaulay ring. The Cohen–Macaulay definition given below is equivalent to that given in Definition 2.1.15.

Now let $(R, \mathfrak{m})$ be a Noetherian local ring, and let $k = R / \mathfrak{m}$.

Definition 7.5.1. The local ring $(R, \mathfrak{m})$ is *Cohen–Macaulay* if $\mathrm{Ext}^i_R(k, R) = 0$ for $i < \dim R$; it is *Gorenstein* if in addition, we have, $\mathrm{Ext}^{\dim R}_R(k, R) \cong k$.

Definition 7.5.2. An algebraic variety X is *Gorenstein at a point* $x \in X$, if the stalk $\mathcal{O}_{X,x}$ is Gorenstein; X is *Gorenstein*, if it is Gorenstein at all $x \in X$. A projective variety $X = \mathrm{Proj}\, S$ is *arithmetically Gorenstein*, if $\hat{X}$ (the cone over X) is Gorenstein.

Remark 7.5.3. Similar to the case with Cohen–Macaulayness (see Lemma 6.1.1), the cone $\hat{X}$ is Gorenstein if and only if it is so at its vertex. In particular, if $\hat{X}$ is Gorenstein, then so is X.

In order to give a combinatorial characterization for Gorenstein Schubert varieties, we use the fact that the homogeneous coordinate rings of Schubert varieties are Hodge algebras, also known as *algebras with straightening laws*, abbreviated as ASL. We give the definition here; we recommend [16, 30] for more details on ASL's.

Let H be a finite partially ordered set and N be the set of nonnegative integers. A *monomial* $\mathcal{M}$ on H is a map from H to N. The *support* of $\mathcal{M}$ is the set $\mathrm{Supp}(\mathcal{M}) := \{x \in H \mid \mathcal{M}(x) \neq 0\}$; $\mathcal{M}$ is *standard* if $\mathrm{Supp}(\mathcal{M})$ is a chain in H (a chain being a totally ordered subset, as in the previous section).

If $\mathcal{R}$ is a commutative ring, and we are given an injection $\varphi : H \hookrightarrow \mathcal{R}$, then to each monomial $\mathcal{M}$ on H we may associate

$$\varphi(\mathcal{M}) := \prod_{x \in H} \varphi(x)^{\mathcal{M}(x)} \in \mathcal{R}.$$

Definition 7.5.4. Let $\mathcal{R}$ be a commutative K-algebra. Suppose that H is a finite partially ordered set with an injection $\varphi : H \hookrightarrow \mathcal{R}$. Then we call $\mathcal{R}$ a *Hodge algebra* or also an *algebra with straightening laws* (abbreviated as ASL) on H over K if the following conditions are satisfied:

ASL-1 The set of standard monomials is a basis of the algebra $\mathcal{R}$ as a K-vector space.

ASL-2 If τ and ϕ in H are incomparable and if

$$\tau\phi = \sum_i a_i \gamma_{i1}\gamma_{i2}\cdots\gamma_{it_i},$$

(where $0 \neq a_i \in K$ and $\gamma_{i1} \leq \gamma_{i2} \leq \cdots \gamma_{it_i}$) is the unique expression for $\tau\phi \in \mathcal{R}$ as a linear combination of distinct standard monomials (guaranteed by ASL-1), then $\gamma_{i1} \leq \tau, \phi$ for every i.

Let X_w be a Schubert variety in $G_{d,n}$, and let $R(w) = K[X_w]$ be its homogeneous coordinate ring. Let $H(w)$ be the Bruhat poset of Schubert subvarieties of X_w (the partial order being given by inclusion).

Proposition 7.5.5. *The homogeneous coordinate ring $R(w)$ for a Schubert variety X_w, $w \in I_{d,n}$, is an ASL (on $H(w)$ over K).*

Proof. The ASL-1 property follows from the discussion following Proposition 5.4.7, while the ASL-2 property follows from Remark 7.2.9. □

Definition 7.5.6. An element z of a lattice $\mathcal{L}$ is called *join irreducible* if $z = x \vee y$ implies $z = x$ or $z = y$. The set of join irreducible elements of $\mathcal{L}$ is denoted by $J(\mathcal{L})$; $J(\mathcal{L})$ inherits the structure of a partially ordered set from $\mathcal{L}$.

Definition 7.5.7. As stated previously, a *chain* is a totally ordered subset of a partially ordered set. We say that a partially ordered set is *ranked* if all maximal chains have the same cardinality.

Let $\mathcal{L}$ be a distributive lattice, and $\mathcal{R}$ a graded ASL domain on $\mathcal{L}$ over a field K. Further, let

$$\deg(\alpha) + \deg(\beta) = \deg(\alpha \vee \beta) + \deg(\alpha \wedge \beta)$$

for all $\alpha, \beta \in \mathcal{L}$. We have the following characterization of the Gorenstein property for $\mathcal{R}$, first a result of [84], but given here as it is stated in [30].

Theorem 7.5.8 (cf. §3 of [30]). *With notation as above, $\mathcal{R}$ is Gorenstein if and only if $J(\mathcal{L})$ is a ranked partially ordered set.*

Using the above Theorem and Proposition 7.5.5, $X(\tau)$ is arithmetically Gorenstein if and only if $J(H(\tau))$ is a ranked partially ordered set.

We will prove an equivalent combinatorial description of arithmetically Gorenstein Schubert varieties, and as a by-product, we obtain an alternate proof of a result of [87, 90]. We begin with the following proposition.

Proposition 7.5.9. *Let $\tau \in I_{d,n}$. Then X_τ is arithmetically Gorenstein if and only if τ consists of intervals $I_1, I_2, \cdots, I_s$ where*

$$I_t = [x_t, y_t], 1 \le t \le s, \; x_{t+1} - y_t = y_t + 2 - x_t, \; 1 \le t \le s-1.$$

(Here, $[x_t, y_t]$ denotes the set $\{x_t, x_{t+1}, \cdots, y_{t-1}, y_t\}$, where it is possible that $x_t = y_t$.)

Proof. We begin the proof with some combinatorial observations. More details can be found in [10]. First of all, it is easily seen that join irreducible elements in $I_{d,n}$ either consist of one segment of consecutive integers, or two segments of consecutive integers in which the first segment begins with 1 (this description of particular elements in $I_{d,n}$ has been seen previously in Theorem 6.4.2). Note that any $\tau \in I_{d,n}$ can be broken into segments $I_1, \ldots, I_s$. If w is a join irreducible element of $I_{d,n}$, $w \le \tau$, then either w is one segment and the first element is less than $\tau_1 (= x_1)$, or w consists of two segments $w = (1, \ldots, t, w_{t+1}, \ldots, w_d)$ such that $w_{t+1} \le \tau_{t+1}$.

The maximal elements of $J(H(\tau))$ are in correspondence with the segments $I_1, \ldots, I_s$ of τ. Let $j_1, \ldots, j_s$ be such that $\tau_{j_1} = x_1, \ldots, \tau_{j_s} = x_s$. Then the maximal elements of $J(H(\tau))$ are $w_1, \ldots, w_s$, where

$$w_1 = [x_1, x_1 + d - 1], \; w_i = ([1, j_i - 1], [x_i, x_i + d - j_i]), \; 2 \le i \le s,$$

(if $x_1 = 1$, then w_1 is not maximal and can be disregarded). The number of elements in a maximal chain of join irreducibles starting at $[1, d]$ and ending at w_i is

$$x_i - (j_i - 1) + (d - j_i) = x_i - 2j_i + d + 1, \; 1 \le i \le s.$$

This is because there are $x_i - (j_i - 1)$ join irreducible elements from w_i to $([1, j_i - 1], [j_i + 1, d + 1])$, and there are $d - j_i + 1$ elements from $([1, j_i - 1], [j_i + 1, d + 1])$ to $[1, d]$ (but we have double counted one point this way).

Thus, $J(H(\tau))$ is ranked if and only if the number of elements in a maximal chain of join irreducibles ending at w_i is the same for $1 \le i \le s$. This is equivalent to the following: for $1 \le i \le s-1$

$$x_i - 2j_i + d + 1 = x_{i+1} - 2j_{i+1} + d + 1,$$
$$x_i + 2j_{i+1} - 2j_i = x_{i+1}.$$

By the setup, we have $y_i - x_i = j_{i+1} - j_i - 1$. Substituting into the above equation, we have

$$x_i + 2(y_i - x_i + 1) = x_{i+1},$$

and the desired result follows for all $1 \le i \le s$. □

To $\tau \in I_{d,n}$, we associate a Young diagram λ^τ (as done previously, e.g., Example 7.4.16) as follows: Let $\tau = (\tau_1, \cdots, \tau_d)$ (as a d-tuple). Set

$$\lambda_r^\tau = \tau_r - r, \ \forall\, 1 \leq r \leq d.$$

Thus we write $\lambda^\tau = (\lambda_d^\tau, \cdots, \lambda_1^\tau)$; when there is no room for confusion, we drop the superscript and just write λ.

Let us write λ as a Young diagram, and place the bottom left corner of the first row (λ_1) at $(0,0)$ on the grid; then each block is a square of unit 1. Thus our diagram will be d-units high and λ_d-units wide.

Definition 7.5.10. The partition λ satisfies the *outer corner condition* if all of the outer corners lie on a line of slope 1; we also refer to this as "the outer corners of λ lie on the same antidiagonal" (same terminology as in [90]).

Example 7.5.11. Let $n = 14, d = 6, \tau = (3,4,5,9,11,12)$, and thus $\lambda^\tau = (6,6,5,2,2,2)$. We write λ^τ as a diagram:

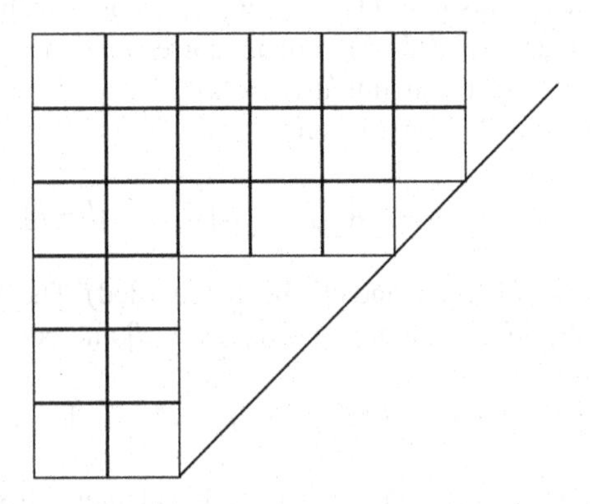

We can see that λ^τ satisfies the outer corner condition. Now let $\tau' = (1,3,4,5,7,10)$, thus $\lambda^{\tau'} = (4,2,1,1,1,0)$:

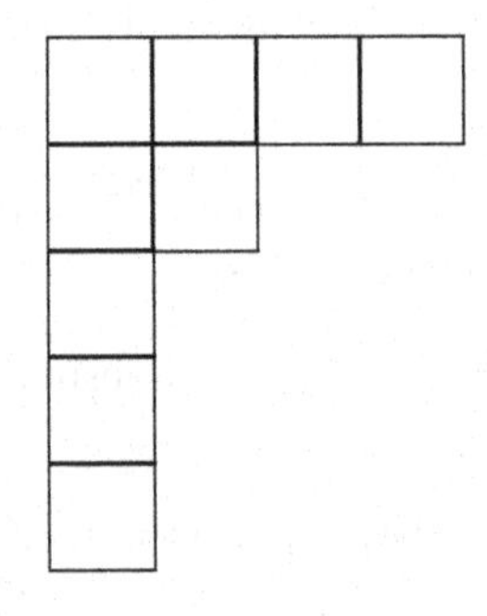

We can see that $\lambda^{\tau'}$ does not satisfy the outer corner condition.

Theorem 7.5.12. *Let $\tau \in I_{d,n}$. Then X_τ is arithmetically Gorenstein if and only if λ^τ satisfies the outer corner condition.*

Proof. We must show that the condition given in Proposition 7.5.9 is equivalent to the outer corner condition. Note that any $\tau \in I_{d,n}$ can be written as a series of intervals $I_1, I_2, \ldots, I_s$, and each interval corresponds to one outer corner in λ^τ. For two successive outer corners to lie on the same antidiagonal, their difference in width must match the height of the lower block. For corners (or "blocks") represented by $I_t = [x_t, y_t]$ and $I_{t+1} = [x_{t+1}, y_{t+1}]$, their difference in width is given by $x_{t+1} - y_t - 1$; while the height of the I_t-block is simply the length of the interval, given by $y_t - x_t + 1$. The result follows. (The reader may want to try this on an example, such as $\tau = (3, 4, 5, 9, 11, 12)$, where $I_1 = [3, 5]$, $I_2 = 9$, $I_3 = [11, 12]$.) □

Example 7.5.13. In Example 6.4.7, we listed the ten Schubert varieties of $G_{3,6}$ that are nonsingular, and hence Gorenstein. Of the remaining ten singular Schubert varieties in $G_{3,6}$, we have the following six that are Gorenstein (the reader is encouraged to check the outer corner condition for each):

$$X_{(135)}, X_{(146)}, X_{(245)}, X_{(236)}, X_{(246)}, X_{(356)}.$$

This leaves only four Schubert varieties in $G_{3,6}$ that are not Gorenstein.

Remark 7.5.14. The non-Gorenstein locus of a Schubert variety can be found and described as a union of Schubert subvarieties, similar to the description of the singular locus for a Schubert variety that is singular, cf. [80].

Part III
Flag Varieties and Related Varieties

Chapter 8
The Flag Variety: Geometric and Representation Theoretic Aspects

In this chapter, we discuss how some of the previous results on the Grassmannian apply to the flag variety. We do not provide proofs for all statements included here. For details, the reader may refer to [45].

8.1 Definitions

Recall from §5.1 that we have a canonical identification of the flag variety $\mathcal{F}l_n$ with G/B, where $G = SL_n$ and B is the Borel subgroup of upper triangular matrices.

Now we define a Schubert variety in G/B for each $w \in S_n$. For a permutation w, let n_w be a permutation matrix of G associated to w; i.e., the only nonzero entry of n_w in column i is in row $w(i)$. The *Schubert variety* $X(w)$ is defined to be the Zariski closure of the B-orbit of n_wB, with the canonical reduced structure:

$$X(w) = \overline{B \cdot n_w B}(mod\, B).$$

The inclusion of the flag variety into the product of Grassmannians (cf. §5.1) induces a partial order on the Schubert varieties of $\mathcal{F}l_n$. Given $w = (i_1 \dots i_n) \in S_n$, where $w(1) = i_1, \dots, w(n) = i_n$, let $\pi_d(w) = (i_1, \dots, i_d) \in I_{d,n}$, where the elements $i_1, \dots, i_d$ are rearranged in increasing order. Then, for $w, w' \in S_n$, we define the partial order

$$w \geq w', \text{ if } \pi_d(w) \geq \pi_d(w')(in\ I_{d,n}),\ \forall\, 1 \leq d \leq n-1.$$

Remark 8.1.1. The above partial order is called *the Bruhat-Chevalley order*.

With the partial order as defined above, we have the same inclusion condition on Schubert varieties in the flag variety as we did in the Grassmannian. Namely,

$$X(w) \supseteq X(w') \Leftrightarrow w \geq w'.$$

V. Lakshmibai, J. Brown, *The Grassmannian Variety*,
Developments in Mathematics 42, DOI 10.1007/978-1-4939-3082-1_8

8.2 Standard Monomials on the Flag Variety

Recall from §5.3 that B is the semidirect product of U and T, where U is the unipotent subgroup of upper triangular matrices (with 1's on the diagonal), and T is the maximal torus of diagonal matrices. Thus the character group of B is equal to the character group of T; i.e., $X(B) = X(T)$. Let $\lambda \in X(B)$; thus $\lambda : B \to K^*$. To λ, we associate a line bundle over G/B. Define $G \times^B K$ as

$$G \times K/\sim \text{ where } (g, x) \sim (gb, \lambda(b)x)\,,\ g \in G,\ b \in B,\ x \in K.$$

We denote the induced line bundle by $L(\lambda)$. Thus, $L(\lambda)$ is the line bundle on G/B, associated to the principal B-bundle $G \to G/B$ for the action of B on K given by λ. We have

$$H^0\,(G/B, L(\lambda)) = \{s : G/B \to G \times^B K \mid \pi \circ s = Id\}$$

where π is the natural projection $\pi : G \times^B K \to G/B$, $\pi(g, x) = gB$, and Id is the identity map on G/B.

Proposition 8.2.1. *We have a natural identification:*

$$H^0\,(G/B, L(\lambda)) \cong \{f \in K[G] \mid f(gb) = \lambda(b)f(g), g \in G, b \in B\}.$$

The mapping

$$\{f \in K[G] \mid f(gb) = \lambda(b)f(g),\ g \in G, b \in B\} \to H^0\,(G/B, L(\lambda))$$

is given by associating $f \in K[G]$ to the map $s : G/B \to G \times^B K$ such that $s(gB) = (g,f(g))$. For the rest of the details of the proof, the reader may refer to [45, Chap. 4, §3].

Recall from §5.2.1, we have $\varepsilon_i \in X(T)$, $\varepsilon_i(t_1, \ldots, t_n) = t_i$, $1 \le i \le n$. Following the notation and terminology of [8], let $\omega_1, \ldots, \omega_{n-1}$ be the *fundamental weights*, defined as

$$\omega_i = \varepsilon_1 + \ldots + \varepsilon_i - \frac{i}{n}\sum_{j=1}^{n} \varepsilon_j, \quad 1 \le i \le n-1.$$

We note that $\omega_i(t_1, \ldots, t_n) = t_1 \cdots t_i$, and thus $\{\omega_1, \ldots, \omega_{n-1}\}$ is a canonical $\mathbb{Z}$-basis of $X(T)$.

Let $\lambda \in X(T)$ be such that $\lambda = a_1\omega_1 + \ldots + a_{n-1}\omega_{n-1}$, where $a_1, \ldots, a_{n-1}$ are nonnegative integers; such a λ is a called a *dominant integral weight*. To such a λ, we associate a Young diagram (which is also denoted by λ) with parts $\lambda_i = a_i + \ldots + a_{n-1}$, $1 \le i \le n-1$.

Example 8.2.2. Let $n = 5$ and $\lambda = 2\omega_1 + 3\omega_2 + \omega_4$. Then the associated Young diagram is $(6, 4, 1, 1)$:

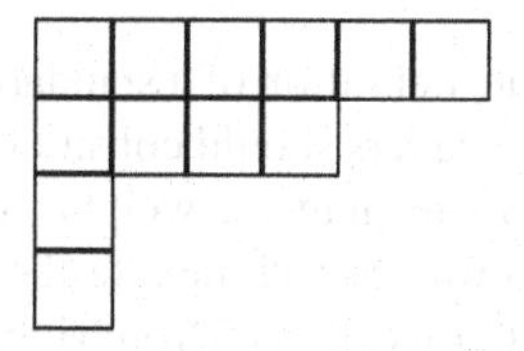

Definition 8.2.3. Given a Young diagram λ, a filling of the diagram with integers is called a *Young tableau* of shape λ. A Young tableau is *standard* if the entries along every row are nondecreasing, and the entries down each column are strictly increasing.

Now let λ be a Young diagram with at most $n - 1$ rows. Let $\mathcal{L}_{n,\lambda}$ be the set of all standard Young tableau of shape λ with entries from $\{1, \ldots, n\}$. Let $\Lambda \in \mathcal{L}_{n,\lambda}$; suppose $\underline{i} = (i_1, \ldots, i_d)$ is some column in Λ, $1 \leq i_1 < \ldots < i_d \leq n$. Set $p_\Lambda = \prod_{\underline{i} \in \Lambda} p_{\underline{i}} \in K[p_\tau, \tau \in I_{d,n}, 1 \leq d \leq n - 1]$ (here, $p_{\underline{i}}$ denotes the Plücker coordinate associated to $\underline{i}$). We say that p_Λ is a *standard monomial on G/B*. Note that $p_\Lambda \in H^0(G/B, L(\lambda))$.

The following theorem indicates that the definition of a standard monomial on G/B is the appropriate generalization of the definition of standard monomials on the Grassmannian. See [45, Chap. 7, Theorem 2.1.1] for a proof.

Theorem 8.2.4. *Let λ be a dominant integral weight (so that the associated Young diagram has at most $n - 1$ rows). Let $\mathcal{L}_{n,\lambda}$ be as above. Then $\{p_\Lambda, \Lambda \in \mathcal{L}_{n,\lambda}\}$ is a basis for $H^0(G/B, L(\lambda))$.*

We now show (by an example) that the definition of a standard monomial on the flag variety, does not generalize to a Schubert variety (in the flag variety). Recall that in the case of the Grassmannian $G_{d,n}$, a standard monomial M on $G_{d,n}$ when restricted to a Schubert variety X_w is either 0 or remains standard on X_w; further, the set of standard monomials on $G_{d,n}$ whose restrictions to X_w are nonzero is linearly independent on X_w (cf. Proposition 5.4.6). This phenomenon does not hold for Schubert varieties in SL_n/B, as illustrated in the following example.

Example 8.2.5. Let $G = SL_3$ and $w = (312) \in S_3$. We first note that $p_{(1)}p_{(2,3)} = p_{(3)}p_{(1,2)} - p_{(2)}p_{(1,3)}$, which can be seen by considering $p_{(i_1,\ldots,i_d)}$ as the $d \times d$-minor of a generic 3×3 matrix X, where the rows are $i_1, \ldots, i_d$ and the columns are $1, \ldots, d$. Let

$$T_1 = \begin{array}{|c|c|}\hline 1 & 3 \\ \hline 2 & \\ \cline{1-1}\end{array}\ , \quad T_2 = \begin{array}{|c|c|}\hline 1 & 2 \\ \hline 3 & \\ \cline{1-1}\end{array}\ .$$

Then $p_{(1)}p_{(2,3)} = p_{T_1} - p_{T_2}$. Now we have that $p_{(2,3)}$ vanishes on $X(w)$, because $X(w)$ projects onto $X_{(1,3)} \subset G_{2,3}$, and thus $p_{T_1} = p_{T_2}$ when restricted to $X(w)$.

We can also see that these monomials are nonzero on $X(w)$. Thus we have a set of monomials $\{p_{T_1}, p_{T_2}\}$ standard on G/B that do not vanish on $X(w)$, but that are linearly dependent on $X(w)$.

We now give the appropriate definition of a standard monomial on $X(w)$ in G/B; for a more detailed treatment, readers should consult [45, Chap. 7].

As above, let λ be a dominant integral weight (so that the associated Young diagram has at most $n-1$ rows). Let m_i denote the number of columns of λ of length i, for $1 \le i \le n-1$. Let us denote (from right to left) the columns of λ of length i by $\tau_{i,1}, \tau_{i,2}, \ldots, \tau_{i,m_i}$. Thus the columns of λ (from right to left) are denoted

$$\tau_{1,1}, \tau_{1,2}, \ldots, \tau_{1,m_1}, \ldots, \tau_{(n-1),1}, \tau_{(n-1),2}, \ldots, \tau_{(n-1),m_{n-1}}.$$

(If $m_i = 0$ for some i, the family $\{\tau_{i,j},\ 1 \le j \le m_i\}$ is understood to be empty.) Let Λ be a standard tableau of shape λ with entries from $\{1, \ldots, n\}$; then a column $\tau_{i,j}$ may be thought of as an element of $I_{i,n}$.

Definition 8.2.6. Let $w \in S_n$. Let Λ be a tableau of shape λ with entries from $\{1, \ldots, n\}$. Λ is a *Young tableau on* $X(w)$ if

$$\pi_i(w) \ge \tau_{i,j}, \quad \text{for all } 1 \le j \le m_i,$$

for all $1 \le i \le n-1$ (π_i is as defined in §8.1); note that $p_\Lambda|_{X_w} \ne 0$. In addition, Λ is a *standard tableau on* $X(w)$ if there exists a sequence

$$\phi_{1,1}, \ldots, \phi_{1,m_1}, \ldots, \phi_{(n-1),1}, \ldots, \phi_{(n-1),m_{n-1}} \in S_n$$

such that

$$X(w) \supseteq X(\phi_{1,1}) \supseteq \cdots \supseteq X(\phi_{1,m_1}) \supseteq X(\phi_{2,1}) \supseteq \cdots \supseteq X(\phi_{(n-1),m_{n-1}}) \qquad (*)$$

and

$$\pi_i(\phi_{i,j}) = \tau_{i,j},\ 1 \le j \le m_i,\ 1 \le i \le n-1. \qquad (**)$$

When this is the case, we say p_Λ is *a standard monomial of type Λ on $X(w)$*.

Remark 8.2.7. In the case that $w = (n\,n-1 \ldots 1)$, we have $X(w) = G/B$, and the definition of Λ being standard on $X(w)$ coincides with the definition given in Definition 8.2.3.

Example 8.2.8. Let us return to Example 8.2.5, where $w = (3, 1, 2) \in S_3$ and

$$T_1 = \begin{array}{|c|c|}\hline 1 & 3 \\ \hline 2 & \\ \cline{1-1}\end{array}\ .$$

Thus $\tau_{1,1} = (3) \in I_{1,3}$ and $\tau_{2,1} = (1,2) \in I_{2,3}$. We set $\phi_{1,1} = (312)$ and $\phi_{2,1} = (123)$ and we have a sequence that satisfies both $(*)$ and $(**)$ of Definition 8.2.6. Therefore $p_{T_1} = p_{(3)}p_{(1,2)}$ is standard on $X(w)$.

On the other hand, we have

$$T_2 = \begin{array}{|c|c|}\hline 1 & 2 \\ \hline 3 & \\ \cline{1-1} \end{array} ,$$

and thus $\tau_{1,1} = (2)$ and $\tau_{2,1} = (1,3)$. Suppose we have a sequence $\phi_{1,1}, \phi_{2,1}$ satisfying $(*)$ and $(**)$. By $(**)$, $\phi_{1,1}$ may be (213) or (231), but to satisfy $(*)$, we must have $w \geq \phi_{1,1}$, and thus $\phi_{1,1}$ should be taken as (213). But for $\phi_{2,1}$ to satisfy $(**)$, we have that $\phi_{2,1}$ must equal (132) or (312), neither of which satisfies $(*)$ (which states $\phi_{1,1} \geq \phi_{2,1}$). Thus we have $p_{T_2} = p_{(2)}p_{(1,3)}$ is not standard on $X(w)$.

Theorem 8.2.9. *Let λ be a dominant integral weight (so that the associated Young diagram has at most $n-1$ rows). Let $w \in S_n$, and let $\mathcal{L}^w_{n,\lambda}$ be the set of Young tableau of shape λ that are standard on $X(w)$. Then $\{p_\Lambda, \Lambda \in \mathcal{L}^w_{n,\lambda}\}$ is a basis for $H^0(X(w), L(\lambda))$.*

See [45, Chap. 7] for a proof of the above theorem in the generalized setting of a union of Schubert varieties.

As a consequence of the above theorem, we have the following vanishing theorem.

Theorem 8.2.10. *Let λ be a dominant integral weight, and X be a union of Schubert varieties in G/B. Then*

$$H^i(X, L(\lambda)) = 0 \textit{ for } i > 0.$$

8.3 Toric Degeneration for the Flag Variety

Let $n \geq 2$, and let

$$H_n = \bigcup_{d=1}^{n-1} I_{d,n}.$$

We define a partial order for H_n. Let $\tau = (i_1, \ldots, i_r)$ and $\phi = (j_1, \ldots, j_s)$. Then

$$\tau \geq \phi, \text{ if } r \leq s \text{ and } i_t \geq j_t,\ 1 \leq t \leq r.$$

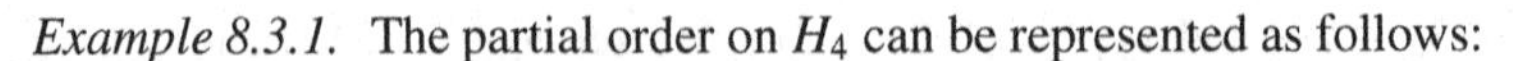

Example 8.3.1. The partial order on H_4 can be represented as follows:

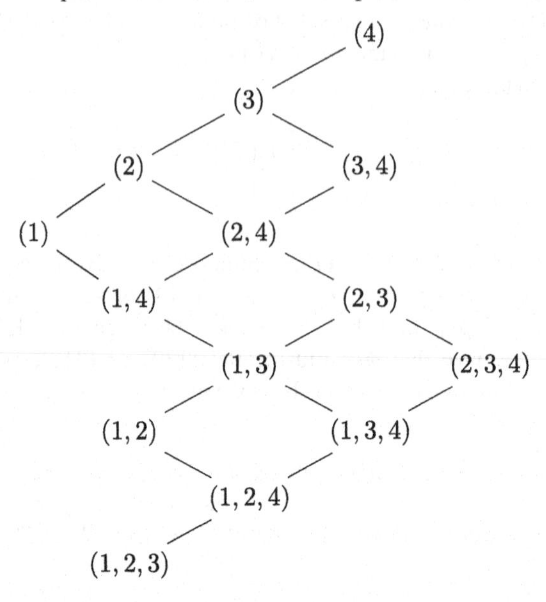

Remark 8.3.2. H_n is a distributive lattice for all $n \geq 2$.

Using the standard monomial basis for $\mathcal{F}l_n$ as described in the previous section and proceeding as in §7.2, we have the following theorem:

Theorem 8.3.3. *The multicone over $\mathcal{F}l_n$, namely,*

$$Spec\left(\bigoplus_{\{\lambda \text{ dominant, integral}\}} H^0(G/B, L(\lambda))\right)$$

flatly degenerates to the Hibi toric variety associated to H_n.

For a proof see [45, Chap. 11, §8].

8.4 Representation Theoretic Aspects

Let $G = SL_n(K)$, and B, T, and $X(T)$ be as in the previous sections. We shall write the elements of $X(T)$ additively. Recall from §8.2 that $X(T) = X(B)$, and that $\lambda \in X(T)$ induces a line bundle on G/B, namely, it is the line bundle on G/B associated to the principal fiber bundle $G \to G/B$, for the action of B on K given by λ. Thus we have a map

$$\theta : X(T) \to \operatorname{Pic} G/B; \quad \lambda \mapsto L(\lambda).$$

(here, Pic G/B is the set of isomorphism classes of line bundles on G/B). By a result of Chevalley (cf. [13]), if G is simply connected, then θ is an isomorphism.

In §8.2, we gave a description of the total space, $G \times^B K$, of $L(\lambda)$. We note that we have a canonical action of G on $G \times^B K$ (given by left multiplication on the first factor), and the map $\pi : G \times^B K \to G/B$ is G-equivariant, i.e., $\pi(h \cdot (g, x)) = h \cdot \pi(g, x), h, g \in G, x \in K$; hence $H^0(G/B, L(\lambda))$ acquires a G-module structure. Below we collect some additional facts concerning $H^0(G/B, L(\lambda))$. The proofs of much of what follows can be found in [34, Chap. 2].

Given a G-module V, we have that as a T-module, V is completely reducible, i.e., V breaks up as a direct sum of T-weight spaces:

$$V = \bigoplus_{\chi \in X(T)} V_\chi,$$

where $V_\chi = \{v \in V \mid t \cdot v = \chi(t) \cdot v, \ \forall\, t \in T\}$. A character $\chi \in X(T)$ is a *highest weight in V* if there exists a nonzero element $v \in V$ such that $b \cdot v = \chi(b) \cdot v$ for all $b \in B$. (Recall that $X(B) = X(T)$.)

(1) We first note that there exists a bijection between the set of dominant integral weights and the set of isomorphism classes of finite dimensional irreducible G-modules (by irreducible, we mean that there is no proper, nonzero G-submodule). Further, for a dominant integral weight λ, denoting the associated irreducible G-module by $V(\lambda)$, we have that λ occurs as the unique highest weight in $V(\lambda)$.

(2) We have that $H^0(G/B, L(\lambda)) \neq 0$ if and only if λ is a dominant integral weight.

Action of S_n on $X(T)$: For $w \in S_n, \lambda \in X(T)$, $w(\lambda)$ is defined to be the character $w(\lambda) : T \to K^*$ given by $w(\lambda)(t) = \lambda(n_w^{-1} t n_w), t \in T, n_w$ being a lift in $N(T)$ (normalizer of T in G) for w; note that $w(\lambda)(t)$ has to be defined as $\lambda(n_w^{-1} t n_w)$, rather than $\lambda(n_w t n_w^{-1})$, in order to have multiplicative property for the induced map $S_n \to Aut(X(T))$. Also note that $w(\lambda)$ is independent of the lift n_w.

(3) Taking $V = H^0(G/B, L(\lambda))$, where λ is a dominant integral weight, there exists a unique B-stable line in $H^0(G/B, L(\lambda))$, with $i(\lambda)$ as the corresponding highest weight, where i is the *Weyl involution* (namely, $i(\lambda) = -w_0(\lambda), w_0 = (n\, n-1 \ldots 1)$ being the element of largest length in S_n, the elements of $X(T)$ being written additively). Further, the G-submodule generated by this line is $V(i(\lambda))$, the irreducible G-module with highest weight $i(\lambda)$.

(4) If λ is a dominant integral weight, then $H^0(G/B, L(\lambda))$ is an indecomposable G-module; i.e., $H^0(G/B, L(\lambda))$ cannot be expressed as a direct sum of nontrivial, proper G-submodules.

(5) If K has characteristic 0, then Fact (4) together with the complete reducibility of G (in characteristic 0) implies that $H^0(G/B, L(\lambda))$ is G-irreducible with highest weight $i(\lambda)$ and conversely ($i(\lambda)$ being the Weyl involution). Thus in characteristic 0, the set

$$\{H^0(G/B, L(\lambda)),\ \lambda \text{ dominant, integral}\}$$

gives all finite dimensional irreducible G-modules, up to isomorphism.

Recall that a line bundle L on an algebraic variety X is *very ample* if there is a closed immersion $i : X \hookrightarrow \mathbb{P}^N$ such that $i^* (\mathcal{O}_{\mathbb{P}^n}(1)) = L$; and L is *ample* if some power of L is very ample.

(6) Taking, $X = G/B$, we have that $L(\lambda)$ is ample if and only if $L(\lambda)$ is very ample, if and only if $\lambda = \sum_{i=1}^{n-1} a_i\omega_i$ where $a_i \in \mathbb{N}, 1 \leq i \leq n-1$. Such a λ is called *regular* and dominant.

Definition 8.4.1. For λ dominant integral, $H^0(G/B, L(\lambda))^*$ is called the *Weyl module*, and $H^0(G/B, L(\lambda))$ is called the *dual Weyl module*, associated to λ.

Definition 8.4.2. For λ dominant integral, and $w \in S_n$, $H^0(X(w), L(\lambda))$ is called the *Demazure module*, associated to λ, w.

8.4.1 Application to $G_{d,n}$

We have, $L(\omega_d) = \mathcal{O}_{G_{d,n}}(1), 1 \leq d \leq n-1$. Let $\lambda = r\omega_d$, $r \in \mathbb{N}$ (a dominant, integral weight). All of the above facts apply to the G-module $H^0(G_{d,n}, L(r\omega_d))$.

8.5 Geometric Aspects

In this section, we state some of the geometric results on Schubert varieties in the flag variety. For details, the reader may refer to [45].

Recall from the previous section that the line bundle $L(\lambda)$ on $X = G/B$ is ample if and only if $L(\lambda)$ is very ample, if and only if $\lambda = \sum_{i=1}^{n-1} a_i\omega_i$ where a_i is a positive integer for $1 \leq i \leq n-1$. Thus λ is regular and dominant. We have the corresponding projective embedding

$$i_\lambda : G/B \hookrightarrow \mathbb{P}(V(\lambda)).$$

This induces an embedding

$$i_\lambda : X(w) \hookrightarrow \mathbb{P}(V(\lambda)), w \in S_n.$$

Note that $V(\lambda) = H^0(G/B, L(\lambda))^*$.

We begin with the following result (cf. [45, Chap. 7, Thm 6.4.1]).

Theorem 8.5.1. *Let $w \in S_n$, λ be a regular, dominant weight, and i_λ the corresponding projective embedding. Let $\widehat{X(w)}$ be the cone over $X(w)$, namely, Spec$\left(\bigoplus_{r\in\mathbb{Z}_+} H^0(X(w), L(r\lambda))\right)$. Then $\widehat{X(w)}$ is Cohen–Macaulay.*

We now discuss singularities of Schubert varieties in G/B.

8.5.1 Description of the tangent space

We have the following description of the tangent space at a T-fixed points e_τ in X_w similar to Theorems 6.5.5, 6.5.7.

Theorem 8.5.2. *Let $w \in S_n$.*

1. *$T_{e_{id}}X_w$ is spanned by $\{X_{ij}, i > j \mid w \geq (i,j)\}$,*
2. *$\dim T_{e_{id}}X_w = \#\{(i,j) \mid w \geq (i,j)\}$.*

Here, (i,j) is the transposition of i,j, and X_{ij} is the elementary matrix with 1 at the ij^{th} place and all other entries equal to 0 (the same notation as used in §6.5).

Corollary 8.5.3. *The Schubert variety X_w is nonsingular if and only if*

$$\dim X_w = \#\{(i,j) \mid w \geq (i,j)\}.$$

Note that for $w = (a_1, \ldots, a_n)$, $\dim X_w = \#\{(i,j), i < j \mid a_i > a_j\}$.

Theorem 8.5.4. *Let X_w be a Schubert variety, and $e_\tau \in X_w$.*

1. *$T_{e_\tau}X_w$ is spanned by $\{X_{\tau(i)\tau(j)}, i > j \mid w \geq (\tau(i), \tau(j))\tau\}$ (where $X_{\tau(i)\tau(j)}$ is the elementary matrix as in Theorem 8.5.2).*
2. *$\dim T_{e_\tau}X_w = \#\{(i,j), \mid w \geq (\tau(i), \tau(j))\tau\}$.*
3. *The Schubert subvariety $X_\tau \subseteq Sing\, X_w$ if and only if $\dim T_{e_\tau}X_w > \dim X_w$.*

8.5.2 Pattern avoidance

We describe below a criterion for smoothness of Schubert varieties in the Flag variety in terms of certain "pattern avoidance." The proof of the following "pattern avoidance" theorem can be found in [45, Chap. 8, Theorem 15.0.1], though it was first proved by Lakshmibai–Sandhya (cf.[52]).

Theorem 8.5.5. *Let $w = (a_1, \ldots, a_n) \in S_n$. Then $X(w)$ is singular if and only if there exist i, j, k, l, $1 \leq i < j < k < l \leq n$ such that either*

1. *$a_k < a_l < a_i < a_j$, or*
2. *$a_l < a_j < a_k < a_i$.*

Example 8.5.6. The two patterns of the previous theorem can be found in SL_4. In fact, the only singular Schubert varieties in SL_4/B are $X(3412)$ and $X(4231)$.

The singular locus of a singular Schubert variety in G/B was conjectured by Lakshmibai–Sandhya (cf. [52]), and proven in [5, 35, 70]. The statement of the result is also in terms of pattern avoidance, though more technical to state than the theorem above, we include it here. Once again, the motivating examples can be found in SL_4/B, where

$$\mathrm{Sing}X(3412) = X(1324) \text{ and } \mathrm{Sing}X(4231) = X(2143).$$

We suggest that the reader keep these two examples in mind while reading the details below.

Let $X(w)$ be a singular Schubert variety of G/B. Thus for $w = (a_1, \ldots, a_n) \in S_n$, there exist i, j, k, l, $1 \leq i < j < k < l \leq n$ such that either $a_k < a_l < a_i < a_j$ or $a_l < a_j < a_k < a_i$.

We define E_w to be a subset of S_n, consisting of τ's defined as follows (depending upon which type of pattern w contains).

1. If there exist $i < j < k < l$ such that $a_k < a_l < a_i < a_j$, and $\tau = (b_1, \ldots, b_n) \leq w$ such that
   a. There exist i', j', k', l', $1 \leq i' < j' < k' < l' \leq n$ such that
   $$b_{i'} = a_k,\ b_{j'} = a_i,\ b_{k'} = a_l,\ b_{l'} = a_j.$$
   b. If τ' is the element obtained from w by replacing a_i, a_j, a_k, a_l respectively by a_k, a_i, a_l, a_j, then $\tau' \leq \tau$.
   c. If w' is obtained from τ by replacing $b_{i'}, b_{j'}, b_{k'}, b_{l'}$ respectively by $b_{j'}, b_{l'}, b_{i'}, b_{k'}$, then $w' \leq w$.

   Then $\tau \in E_w$.
2. If there exist $i < j < k < l$ such that $a_l < a_j < a_k < a_i$ and $\tau = (b_1, \ldots, b_n) \leq w$ such that
   a. There exist i', j', k', l', $1 \leq i' < j' < k' < l' \leq n$ such that
   $$b_{i'} = a_j,\ b_{j'} = a_l,\ b_{k'} = a_i,\ b_{l'} = a_k.$$
   b. If τ' is the element obtained from w by replacing a_i, a_j, a_k, a_l respectively by a_j, a_l, a_i, a_k, then $\tau' \leq \tau$.
   c. If w' is the element obtained from τ by replacing $b_{i'}, b_{j'}, b_{k'}, b_{l'}$ respectively by $b_{k'}, b_{i'}, b_{l'}, b_{j'}$, then $w' \leq w$.

   Then $\tau \in E_w$.

For any $w \in S_n$, let P_w (respectively, Q_w) be the maximal element of parabolic subgroups of G which leave $\overline{BwB}$ stable under multiplication on the left (resp. right). Let

$$F_w = \{\tau \in E_w \mid P_w \subseteq P_\tau,\ Q_w \subseteq Q_\tau\}.$$

We can now state the singular locus of $X(w)$.

Theorem 8.5.7. *With notation as above,*

$$\mathit{Sing}X(w) = \bigcup_{\tau \in F_w} X(\tau),$$

(where we only need τ maximal in F_w under the Bruhat order).

Example 8.5.8. Let $w = (351624) \in S_6$. Note that w has two patterns of the form $a_k < a_l < a_i < a_j$: $(3, 5, 1, 2)$ and $(5, 6, 2, 4)$. Using the description above, we have

$$\mathrm{Sing}X(w) = X(132654) \cup X(321546).$$

Example 8.5.9. Let $w = (452316) \in S_6$. Note that w has one pattern of the form $a_k < a_l < a_i < a_j$, that is, $(4, 5, 2, 3)$; and w has two patterns of the form $a_l < a_j < a_k < a_i$: $(4, 2, 3, 1)$ and $(5, 2, 3, 1)$. We have

$$\mathrm{Sing}X(w) = X(243516) \cup X(251436) \cup X(421536).$$

Chapter 9
Relationship to Classical Invariant Theory

In this chapter, we describe a connection between classical invariant theory and the Grassmannian variety. Namely, for $G = SL_d(K)$, X the space of $n \times d$ matrices ($n > d$), and $R = K[x_{ij} \mid 1 \leq i \leq n, 1 \leq j \leq d]$ ($X = \text{Spec}(R)$), we have a G action on X by right multiplication, and hence a G action on R. We will show that the categorical quotient $X /\!/ G$ is isomorphic to the cone over the Grassmannian, and thus obtain a K-basis for R^G, the ring of invariants, consisting of standard monomials. In this chapter, we shall work just with the closed points of an algebraic variety.

9.1 Basic Definitions in Geometric Invariant Theory

First, we recall "Jordan-like decomposition" for elements in an algebraic group G over an algebraically closed field K. For details, the reader may refer to [6].

Let V be a finite dimensional vector space over K. An element $r \in \text{End}(V)$ is *nilpotent* if there exists $n \in \mathbb{N}$ such that $r^n = 0$; r is *unipotent* if $r - 1$ is nilpotent (where 1 represents the identity map on V); r is *semisimple* if V has a basis consisting of eigenvectors of r.

We have the following celebrated *Jordan Decomposition*: Let $r \in \text{End}(V)$.

1. **Additive Jordan Decomposition:** There exist unique $r_s, r_n \in \text{End}(V)$ such that r_s is semisimple, r_n is nilpotent, $r_s r_n = r_n r_s$, and such that $r = r_s + r_n$.
2. **Multiplicative Jordan Decomposition:** If r is invertible, then there exist unique $r_s, r_u \in \text{GL}(V)$ such that r_s is semisimple, r_u is unipotent, and $r = r_s r_u = r_u r_s$.

These results generalize to infinite dimensional vector spaces as well, described as follows.

Let V be an infinite dimensional vector space over K. An element $r \in \text{End}(V)$ is *locally finite* if V is spanned by finite dimensional subspaces stable under r. A locally finite $r \in \text{End}(V)$ is *locally nilpotent* (resp. *locally unipotent*, resp.

V. Lakshmibai, J. Brown, *The Grassmannian Variety*,
Developments in Mathematics 42, DOI 10.1007/978-1-4939-3082-1_9

locally semisimple), if its restriction to each finite dimensional r-stable subspace is nilpotent (resp. unipotent, resp. semisimple). The uniqueness of the above Jordan Decomposition (for finite dimensional vector spaces) gives the following:

Let $r \in \mathrm{End}(V)$ be locally finite.

1. **Additive Jordan Decomposition:** There exist unique $r_s, r_n \in \mathrm{End}(V)$ such that r_s is locally semisimple, r_n is locally nilpotent, $r_s r_n = r_n r_s$, and such that $r = r_s + r_n$, inducing the usual Additive Jordan Decomposition on finite dimensional r-stable subspaces.
2. **Multiplicative Jordan Decomposition:** If in addition r is invertible, then there exist unique $r_s, r_u \in \mathrm{GL}(V)$ such that r_s is locally semisimple, r_u is locally unipotent, and $r = r_s r_u = r_u r_s$, inducing the usual Multiplicative Jordan Decomposition on finite dimensional r-stable subspaces.

Let now G be an algebraic group. For $g \in G$, the automorphism $G \cong G$, $x \mapsto xg$ induces a linear automorphism $\rho_g : K[G] \cong K[G]$, $(\rho_g(f))(x) = f(xg), f \in K[G]$, $x \in G$. We have that as an element of $\mathrm{End}(K[G])$, ρ_g is locally finite. We have the following theorem (cf. [6]).

Theorem 9.1.1. *Let $g \in G$. There is a unique factorization $g = g_s g_u$ in G such that $\rho_g = \rho_{g_s}\rho_{g_u}$ is the Multiplicative Jordan Decomposition of ρ_g*

Semisimple and unipotent elements in *G*. With notation as above, we define G_s (resp. G_u), the set of semisimple (resp. unipotent) elements in G as

$$G_s = \{g \in G \mid g = g_s\},\ G_u = \{g \in G \mid g = g_u\}.$$

9.1.1 Reductive Groups

Recall from Definition 5.1.1 that B is a Borel subgroup of G if B is a maximal, connected, solvable subgroup of G. The radical of G, denoted $\mathcal{R}(G)$, is the connected component (through the identity element) of the intersection of all Borel subgroups of G, i.e.,

$$\mathcal{R}(G) = \left(\bigcap_{B \in \mathcal{B}} B\right)^{\circ},$$

where $\mathcal{B}$ is the set of all Borel subgroups of G.

Definition 9.1.2. An algebraic group G is *reductive* if $\mathcal{R}(G)_u = \{e\}$, e being the identity element.

Remark 9.1.3. The algebraic groups $GL_d(K)$ and $SL_d(K)$ are reductive.

9.2 Categorical Quotient

In this section, we recall some generalities on Geometric Invariant Theory. For details, the reader may refer to [51, 76, 78].

Definition 9.2.1. Let X be an algebraic variety with an action by an algebraic group G. The pair (Y, ϕ), where Y is an algebraic variety and $\phi : X \to Y$ is a morphism of varieties, is a *categorical quotient* if

1. ϕ is G-invariant; i.e., ϕ is constant on G-orbits in X.
2. ϕ has the universal mapping property; i.e., given $f : X \to Z$, where f is constant on G-orbits, there exists a unique $g : Y \to Z$ such that $f = g \circ \phi$:

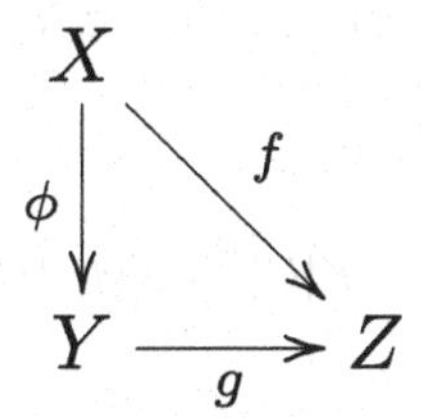

One can see that if a categorical quotient exists, it is unique up to isomorphism. We will denote categorical quotient by $X /\!\!/ G$ (if G acts on the right; a similar notation if G acts on the left).

Definition 9.2.2. Let X be an affine variety, and G a reductive group. We say G *acts linearly* on X if we have an action of G on $\mathbb{A}^r$ such that each $g \in G$ induces a linear map on $\mathbb{A}^r$, and we have a G-equivariant closed immersion $X \hookrightarrow \mathbb{A}^r$.

Example 9.2.3. Let $G = SL_d(K)$, and X be the affine variety of $n \times d$ matrices, thus $X = \mathbb{A}^{nd}$. Then G acts linearly on X by right multiplication.

Definition 9.2.4. Let X be an affine variety, and G a reductive group acting linearly on X.

1. A point $x \in X$ is *semistable* if 0 is not in the orbit closure of x, i.e., $0 \notin \overline{G \cdot x}$. Let X^{ss} denote the subset of semistable points in X.
2. A point $x \in X$ is *stable* if the orbit $G \cdot x$ of x is closed, and the dimension of $G \cdot x$ is equal to the dimension of G. Let X^s denote the subset of stable points in X.

Let $X = \operatorname{Spec} R$ and G be a reductive group acting linearly (on the right) on X. We have an induced action of G on R: for $f \in R$, $(g \cdot f)(x) = f(x \cdot g)$ for $g \in G$, $x \in X$.

We have the inclusion $R^G \hookrightarrow R$, where

$$R^G = \{f \in R \mid g \cdot f = f, \ \forall g \in G\}.$$

We have (thanks to Mumford's conjecture proved by Haboush; for details, the reader may refer to [51, §9.1]) that R^G is a finitely generated K-algebra. Let $Y = \operatorname{Spec}(R^G)$, and let ϕ be the morphism $\phi : X \to Y$ induced by the inclusion above.

Remark 9.2.5. Note that $\phi : X \to Y$ is such that for $f \in R^G$ and $x \in X$,

$$f(\phi(x)) = f(x),$$

since $\phi^*(f) = f(\in R)$.

We will see that in fact (Y, ϕ) is the categorical quotient $X /\!/ G$.

Lemma 9.2.6. *Given the notation above, we have*

1. *ϕ is surjective.*
2. *ϕ is G-invariant, i.e.,*

$$\phi(x \cdot g) = \phi(x),\ x \in X,\ g \in G.$$

3. *Given an open set $U \subset Y$,*

$$\phi^* : K[U] \to K[\phi^{-1}(U)]^G$$

is an isomorphism.

Proof. Claim. (1): Let y be a (closed) point in Y. Let $\mathbf{m}_y$ be the corresponding maximal ideal in $K[Y](= R^G)$; further let $\{f_1, \ldots, f_t\}$ be generators for $\mathbf{m}_y$. We will find a point $x \in X$ such that $\phi(x) = y$. Consider $\sum_{i=1}^t f_i R$. By [78, Lemma 3.4.2], we have that if $f \in \left(\sum_{i=1}^t f_i R\right) \cap R^G$, then there exists some $s \in \mathbb{N}$ such that $f^s \in \mathbf{m}_y$; this implies (since $\mathbf{m}_y$ is maximal) that $f \in \mathbf{m}_y$. Therefore, $\left(\sum_{i=1}^t f_i R\right) \cap R^G \neq R^G$, and hence $\sum_{i=1}^t f_i R \neq R$. Let $\mathbf{m}_x$ be a maximal ideal of R containing $\{f_1, \ldots, f_t\}$. Note that for $x \in X$ corresponding to $\mathbf{m}_x$, we have $f_i(x) = 0$ for all $i = 1, \ldots, t$. It follows that $\phi(x) = y$, and (1) follows.

Claim. (2): To show $\phi(x \cdot g) = \phi(x)$, $x \in X$, $g \in G$, we should show that

$$(*) \qquad f(\phi(x \cdot g)) = f(\phi(x)),\ \forall f \in K[Y]\ (= R^G),\ x \in X,\ g \in G.$$

Now for $f \in R^G, g \in G, x \in X$, we have

$$f(x \cdot g) = f(\phi(x \cdot g)),\ f(x) = f(\phi(x)),\ f(x \cdot g) = (g \cdot f)(x) = f(x).$$

From this and $(*)$, the G-invariance of ϕ follows.

Claim. (3): For $f \in R^G$, let $Y_f = \{y \in Y \mid f(y) \neq 0\}$. It suffices to prove the result for $U = Y_f$, since the set $\{Y_f \mid f \in R^G\}$ (the set of principal open subsets in Y), is a base for the Zariski topology on Y. Let $X_f = \phi^{-1}(U)$. We have $K[U] = R_f^G$. On the other hand, $K[X_f] = R_f$. It is seen easily that the localization of the G-invariants equals the G-invariants of the localization, i.e., $(R^G)_f = (R_f)^G$. Therefore $K[U] = K[X_f]^G$, and the result follows. □

We now turn our attention to the connection between a certain $SL_d(K)$-action and Grassmannian $G_{d,n}$. Let $X = M_{n,d}$ be the space of $n \times d$ matrices with entries in K, where $n > d$. Let $R = K[x_{ij}, 1 \leq i \leq n, 1 \leq j \leq d]$, where R is the coordinate ring of X (i.e., $X = \text{Spec}R$). Let $G = SL_d(K)$, and let G act on X by right multiplication, and hence G acts on R, where for $g \in G, f \in R$, and $A \in X$,

$$(g \cdot f)(A) = f(Ag).$$

As above, let $Y = \text{Spec}(R^G)$, and let $\phi : X \to Y$ be the morphism induced by the inclusion $R^G \hookrightarrow R$.

For $\underline{i} \in I_{d,n}$, let $f_{\underline{i}} \in R$ be defined such that

$$f_{\underline{i}}(A) = \det A_{\underline{i}},\ A \in X,$$

where $A_{\underline{i}}$ is the $d \times d$ minor of A with rows indexed by $i_1, \ldots, i_d$ (as in §5.2.1).

Note that for $\underline{i} \in I_{d,n}$, $g \in G$, and $A \in X$, we have

$$(g \cdot f_{\underline{i}})(A) = f_{\underline{i}}(Ag) = \det(A_{\underline{i}})\det(g) = \det(A_{\underline{i}}) = f_{\underline{i}}(A),$$

and therefore $f_{\underline{i}} \in R^G (= \{f \in R \mid g \cdot f = f,\ \forall\, g \in G\})$.

Let $X^o \subseteq X$ be the set of $n \times d$ matrices of maximal rank (namely, rank d). In the next two lemmas, we show that $X^o = X^{ss} = X^s$.

Lemma 9.2.7. *Using the notation above,*

$$X^o = X^{ss}.$$

Proof. Let $A \in X \setminus X^o$; thus rank$A = r < d$. There exists a $g \in GL_d(K)$ such that Ag is of the form

$$Ag = \begin{bmatrix} m_{1,1} & \cdots & m_{1,r} & 0 & \cdots & 0 \\ \vdots & \ddots & \vdots & \vdots & \ddots & \vdots \\ m_{n,1} & \cdots & m_{n,r} & 0 & \cdots & 0 \end{bmatrix} = \begin{bmatrix} M_{n,r} & 0_{n,(d-r)} \end{bmatrix},$$

where the columns of $M_{n,r}$ are linearly independent. Let $c = \det(g)$, and let α be a dth root of c. Thus $h = \frac{1}{\alpha} g \in SL_d(K)$. Since $\frac{1}{\alpha} Ag = Ah$, we have that $\frac{1}{\alpha} Ag$ is in the orbit of A; replacing A by $Ah (\in O(A))$, we may suppose

$$A = \frac{1}{\alpha}\begin{bmatrix} M_{n,r} & 0_{n,(d-r)} \end{bmatrix}.$$

Choose integers $a_1, \ldots, a_{d-r}$ such that $r + \sum_{i=1}^{d-r} a_i = 0$, and consider the one-parameter subgroup Γ consisting of $\{D_t, t \neq 0\}$ (i.e., $\Gamma : K^* \to T, \Gamma(t) = D_t$), where

$$D_t = \left\{ \begin{bmatrix} t & 0 & \cdots & 0 & 0 & \cdots & 0 \\ 0 & t & \cdots & 0 & 0 & \cdots & 0 \\ \vdots & \vdots & \ddots & \vdots & \vdots & \ddots & \vdots \\ 0 & 0 & \cdots & t & 0 & \cdots & 0 \\ 0 & 0 & \cdots & 0 & t^{a_1} & \cdots & 0 \\ \vdots & \vdots & \ddots & \vdots & \vdots & \ddots & \vdots \\ 0 & 0 & \cdots & 0 & 0 & \cdots & t^{a_{d-r}} \end{bmatrix} \mid t \neq 0 \right\}$$

(where the top left corner is $tId_{r \times r}$). Then

$$AD_t = \frac{1}{\alpha} \begin{bmatrix} tm_{1,1} & \cdots & tm_{1,r} & 0 & \cdots & 0 \\ \vdots & \ddots & \vdots & \vdots & \ddots & \vdots \\ tm_{n,1} & \cdots & tm_{n,r} & 0 & \cdots & 0 \end{bmatrix}.$$

Hence we obtain

$$\lim_{t \to 0}(AD_t) = 0.$$

Thus the origin O is in the closure of $O(A)$, and therefore A is not semistable. Thus we have proven the inclusion

$$X^{ss} \subseteq X^o.$$

We now show the reverse inclusion. Let $A \in X^o$; then there exists some $\underline{i} \in I_{d,n}$ such that $f_{\underline{i}}(A) \neq 0$. Since $f_{\underline{i}} \in R^G$, we have that $f_{\underline{i}}(Ag) = f_{\underline{i}}(A)$ for every Ag in the orbit of A. Thus, $f_{\underline{i}}(B) = f_{\underline{i}}(A) \neq 0$ for every B in the orbit closure of A, implying that 0 is not in the orbit closure. Therefore $A \in X^{ss}$, and the result follows. □

Lemma 9.2.8.

$$X^o = X^s.$$

Proof. Let $A \in X^o$. Suppose $g \in G$ is in the stabilizer of A; i.e., $Ag = A$. Let $\underline{i} \in I_{d,n}$ be such that $\det A_{\underline{i}} \neq 0$. We have that $A_{\underline{i}}$ is an invertible $d \times d$ matrix, and $A_{\underline{i}}g = A_{\underline{i}}$. Therefore, g is the identity matrix. Letting $O(A)$ denote the orbit of A, we have that $O(A)$ is isomorphic to G, and therefore

$$\dim O(A) = \dim G, \text{ for all } A \in X^o.$$

We will now show that for $A \in X^o$, $O(A)$ is closed. Suppose not, let $B \in \overline{O(A)} \setminus O(A)$. By the closed orbit lemma (Lemma 6.3.5), we have that $\dim O(B) < \dim O(A) = \dim G$; thus by the discussion above, $B \notin X^o$. By Lemma 9.2.7, $B \notin X^{ss}$, and thus $0 \in \overline{O(B)} \subseteq \overline{O(A)}$, a contradiction of the fact that $A \in X^o = X^{ss}$. Therefore $O(A)$ is closed.

Thus it follows that if A has maximal rank, then A is stable; i.e., $X^o \subseteq X^s$. This together with Lemma 9.2.7 and the inclusion $X^s \subseteq X^{ss}$ implies that $X^o = X^s$. □

We now return to proving some properties of $\phi : X \to Y$, in the case $X = M_{n,d}$.

Lemma 9.2.9. *If W_1 and W_2 are two disjoint, G-stable, closed subsets of X, then $\phi(W_1)$ and $\phi(W_2)$ are disjoint.*

Proof. Let W_1 and W_2 be disjoint, G-stable, closed subsets of X. First, suppose $W_1 \not\subseteq X^o$. Then by Lemma 9.2.7, we have $0 \in W_1$, and for W_1 and W_2 to be disjoint, we must have $W_2 \subseteq X^o$. Note that since $f_{\underline{i}} \in R^G$, $\underline{i} \in I_{d,n}$, we have the following: for $x_1 \in X \setminus X^o$ we have

$$f_{\underline{i}}(\phi(x_1)) = f_{\underline{i}}(x_1) = 0, \quad \forall\, \underline{i} \in I_{d,n}.$$

On the other hand, for $x_2 \in X^o$, there exists some $\underline{i} \in I_{d,n}$ such that

$$f_{\underline{i}}(\phi(x_2)) = f_{\underline{i}}(x_2) \neq 0.$$

Therefore, $\phi(x_1) \neq \phi(x_2)$. Thus, it suffices to prove the result for $W_1,\ W_2 \subseteq X^o$.

Suppose $y \in \phi(W_1) \cap \phi(W_2)$. Let $A_i \in W_i$ such that $\phi(A_i) = y$, $i = 1, 2$. Since $f_{\underline{i}}(\phi(x)) = f_{\underline{i}}(x)$ for all $x \in X$, $\underline{i} \in I_{d,n}$, we have

$$f_{\underline{i}}(A_1) = f_{\underline{i}}(A_2), \quad \forall\, \underline{i} \in I_{d,n}.$$

Let $\underline{j} \in I_{d,n}$ be such that $f_{\underline{j}}(A_1) = f_{\underline{j}}(A_2) = c \neq 0$ (which we know exists since $A_1, A_2 \in X^o$). Let α be a dth root of c, then

$$A_1 = \alpha g_1,\ A_2 = \alpha g_2,\ g_1, g_2 \in G.$$

Let $B_i = A_i g_i^{-1}$ for $i = 1, 2$, where $B_i \in W_i$ since W_i is G-stable. We also note that B_1, B_2 are such that

$$(B_1)_{\underline{j}} = (B_2)_{\underline{j}} = \alpha I = \begin{bmatrix} \alpha & 0 & \cdots & 0 \\ 0 & \alpha & \cdots & 0 \\ \vdots & \vdots & \ddots & \vdots \\ 0 & 0 & \cdots & \alpha \end{bmatrix}.$$

It is also clear that $f_{\underline{i}}(B_1) = f_{\underline{i}}(B_2)$ for all $\underline{i} \in I_{d,n}$, since $f_{\underline{i}} \in R^G$. Let us use the notation $B_1 = (b^1_{ij})$ and $B_2 = (b^2_{ij})$.

Let k be such that $k \in \{1, \ldots, n\} \setminus \{j_1, \ldots, j_d\}$. For $1 \leq i \leq d$, let $\tau = \{j_1, \ldots, j_{i-1}, k, j_{i+1}, \ldots, j_d\} \in I_{d,n}$ (where we re-order the entries of τ to be in increasing order, if necessary). Then $(B_1)_\tau$ (resp. $(B_2)_\tau$) is obtained by replacing the ith row of the matrix $\alpha I_{d\times d}$ with the kth row of B_1 (resp. B_2). Therefore

$$f_\tau(B_1) = \pm\alpha^{d-1} b^1_{ki}, \quad f_\tau(B_2) = \pm\alpha^{d-1} b^2_{ki},$$

(where the sign is determined by the re-ordering of τ to be in increasing order, and thus will be the same for B_1 and B_2). Therefore, $b^1_{ki} = b^2_{ki}$. From this we obtain that $B_1 = B_2$, implying that W_1 and W_2 are not disjoint, a contradiction. The result now follows. □

Theorem 9.2.10. *Let X, Y, and ϕ be as above, and let Z be an algebraic variety with a G-invariant morphism $F : X \to Z$. Then there exists a unique morphism $\chi : Y \to Z$ such that the following diagram is commutative*

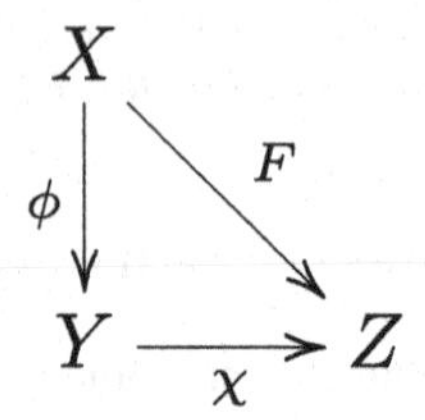

(i.e., ϕ has the universal mapping property).

Proof. We first show uniqueness. Let χ_1 and χ_2 be two such morphisms. For $y \in Y$, since ϕ is surjective (by Lemma 9.2.6), there exists $x \in X$ such that $\phi(x) = y$. Thus

$$\chi_i(y) = \chi_i(\phi(x)) = F(x), \text{ for } i = 1, 2.$$

Thus $\chi_1(y) = \chi_2(y)$ for all $y \in Y$, and uniqueness follows.

We now show that such a morphism χ exists. Let $Z = \bigcup V_i$ be an affine cover, and let $U_i = \{y \in Y \mid F(\phi^{-1}(y)) \in V_i\}$. Denote the restriction of ϕ and F to $\phi^{-1}(U_i)$ by ϕ_i and F_i, respectively. Now we define $\chi_i : U_i \to V_i$ such that $\chi_i(y) = F_i(\phi_i^{-1}(y))$.

We first show that χ_i is well defined, by showing that F is constant on $\phi^{-1}(y)$. Suppose $x_1, x_2 \in \phi^{-1}(y)$, and $z_1 = F(x_1)$, $z_2 = F(x_2)$. Assume $z_1 \neq z_2$. Let $W_1 = F^{-1}(z_1)$, $W_2 = F^{-1}(z_2)$; then W_1 and W_2 are disjoint, G-stable closed subsets of X. By Lemma 9.2.9, $\phi(W_1)$ and $\phi(W_2)$ are also disjoint. But we have $y \in \phi(W_1) \cap \phi(W_2)$ (note that $y = \phi(x_i) \in \phi(W_i), i = 1, 2$), a contradiction. Therefore, $z_1 = z_2$, and F is constant on $\phi^{-1}(y)$, for $y \in Y$.

We now show that χ_i is a morphism of varieties by showing that $\chi_i^* : K[V_i] \to K[U_i]$ is a K-algebra homomorphism. (This shows that χ_i is a morphism in view of the fact that if U and V are two varieties with V affine, we have a bijection $\mathrm{Mor}(U, V) \leftrightarrow \mathrm{Hom}_K(K[V], K[U])$, see [75].) Since F_i is G-invariant, we have

$$F_i^* : K[V_i] \to K[F_i^{-1}(V_i)]^G.$$

By Lemma 9.2.6, we have an isomorphism $\phi_i^* : K[U_i] \cong K[\phi_i^{-1}(U_i)]^G$. We have

$$K[U_i] \cong K[\phi_i^{-1}(U_i)]^G = K[\phi_i^{-1}\chi_i^{-1}(V_i)]^G = K[F_i^{-1}(V_i)]^G.$$

Thus we get that $(\phi_i^*)^{-1} \circ F_i^*$ does indeed induce a K-algebra homomorphism $\chi_i^* : K[V_i] \to K[U_i]$.

Finally, based on the definition of χ_i, it can be easily seen that χ_i and χ_j agree on $U_i \cap U_j$, and hence $\{\chi_i\}$ may be glued to define $\chi : Y \to Z$. □

Since ϕ is surjective and has the universal mapping property, we have the following corollary.

Corollary 9.2.11. *Let $X = M_{n,d}$, and R such that $X = SpecR$. For the right action of $G = SL_d(K)$ on X, we have that $(Spec(R^G), \phi)$ is the categorical quotient $X /\!/ G$.*

Remark 9.2.12. In fact, for any algebraic variety $X = \mathrm{Spec}(R)$ acted on by a reductive group G, it is true that $X /\!/ G = \mathrm{Spec}\left(R^G\right)$. For proof of this, see, for example, [51, §9.2–9.3].

9.3 Connection to the Grassmannian

To begin this section, let us return to the general setup in which $X = \mathrm{Spec}\, R$ is an algebraic variety and G is a reductive group acting on X. Further, let R be a graded K-algebra. Let $f_1, \ldots, f_N$ be homogeneous G-invariant elements in R; i.e., $\{f_1, \ldots f_N\} \subseteq R^G$. Let $S = K[f_1, \ldots, f_N]$. In the remaining discussion, we find sufficient conditions to conclude $S = R^G$, and then apply this to the Grassmannian case.

We begin by stating Zariski's Main Theorem (cf. [75, III. §9]).

Theorem 9.3.1. *Let $F : X \to Y$ be a morphism of algebraic varieties X and Y such that*

1. *F is surjective,*
2. *fibers of F are finite,*
3. *F is birational,*
4. *Y is normal.*

Then F is an isomorphism.

(Recall that a morphism $\phi : X \to Y$ be of algebraic varieties is *dominant* if $\overline{\phi(X)} = Y$. If in addition, the induced map $\phi^* : K(Y) \hookrightarrow K(X)$ of rational functions is an isomorphism, then ϕ is *birational.*)

We shall verify the hypotheses in Zariski's Main Theorem for the morphism $\xi : \mathrm{Spec}(R^G) \to \mathrm{Spec}(S)$, induced by the inclusion $S \subseteq R^G$. First, we give a sufficient condition for the fibers of ξ to be finite.

Lemma 9.3.2. *With notation as above, let $f_1, \ldots, f_N \in R^G$ be homogeneous such that for any $x \in X^{ss}$, we have $f_i(x) \neq 0$ for at least one i, $1 \leq i \leq N$. Then the fibers of $Spec(R^G) \to Spec(S)$ are finite.*

Proof. We break the proof of this lemma into the following two cases:

Case 1. Let $f_1, \ldots, f_N$ be homogeneous of the same degree, say d. Let $Y = \mathrm{Spec}(R^G)$ (the categorical quotient $X /\!/ G$ by Remark 9.2.12), and let us denote the categorical quotient map by $\phi : X \to Y$. Let $X_1 = \mathrm{Proj}(R)$. We define the semistable points of X_1 to be

$$X_1^{ss} = \{x \in X_1 \mid \text{there exists } \hat{x} \in X^{ss} \text{ such that } \hat{x} \text{ lies over } x\}.$$

Let $Y_1 = \mathrm{Proj}(R^G)$; we have (see,for example, [51, §9.6]) $Y_1 = X_1^{ss} /\!\!/ G$. We let $\phi_1 : X_1^{ss} \to Y_1$ be the canonical quotient map. Let $\psi : X \to \mathbb{A}^N$, $x \mapsto (f_1(x), \ldots, f_N(x))$. Since $f_1, \ldots, f_n$ are G-invariant, and since ϕ is the canonical quotient map, we get an induced map $\rho : Y \to \mathbb{A}^N$, forming the following commutative diagram:

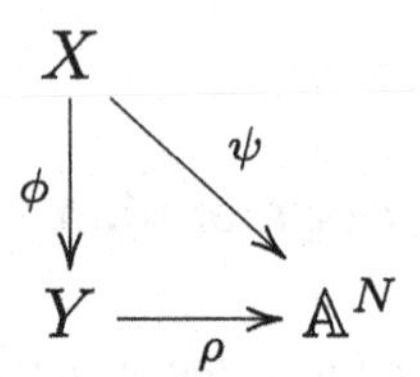

Note that the image of ψ is equal to Spec(S), and thus if we show that ρ is a finite morphism, the result will follow. Now the diagram above induces the following commutative diagram:

$$\begin{array}{ccc} X_1^{ss} & & \\ {\scriptstyle \phi_1}\downarrow & \searrow^{\psi_1} & \\ Y_1 & \xrightarrow[\rho_1]{} & \mathbb{P}^{N-1} \end{array}$$

Note that $\psi_1 : X_1^{ss} \to \mathbb{P}^{N-1}$ is defined in view of the hypothesis that for any $x \in X^{ss}, f_i(x) \neq 0$ for at least one i.

Before we show that ρ is a finite morphism, we show that ρ_1 is a finite morphism. We have that $f_1, \ldots, f_N$ are sections of the ample line bundle $\mathcal{O}_{X_1}(d)$, and that this line bundle descends to an ample line bundle on Y_1, which we denote by $\mathcal{O}_{Y_1}(d)$. Since $f_1, \ldots, f_N$ are G-invariant, we have $\{f_1, \ldots, f_N\} \subseteq H^0(Y_1, \mathcal{O}_{Y_1}(d))$, and thus

$$\rho_1^* (\mathcal{O}_{\mathbb{P}^{N-1}}(1)) = \mathcal{O}_{Y_1}(d).$$

This implies $\rho_1^* (\mathcal{O}_{\mathbb{P}^{N-1}}(1))$ is ample, and hence ρ_1 has finite fibers (note that the restriction of $\rho_1^* (\mathcal{O}_{\mathbb{P}^{N-1}}(1))$ to any fiber of ρ_1 is ample and trivial, and thus the dimension of any fiber is zero).

Since ρ_1 is finite, $\mathcal{O}_{Y_1}$ is a coherent $\mathcal{O}_{\mathbb{P}^{N-1}}$-module. We have that for $n \in \mathbb{N}$,

$$H^0(\mathbb{P}^{N-1}, \mathcal{O}_{Y_1} \otimes \mathcal{O}_{\mathbb{P}^{N-1}}(n)) \cong H^0(Y_1, \rho_1^* (\mathcal{O}_{\mathbb{P}^{N-1}}(n)),$$

because the direct image of $\rho_1^* (\mathcal{O}_{\mathbb{P}^{N-1}}(n))$ by ρ_1 is $\mathcal{O}_{Y_1} \otimes \mathcal{O}_{\mathbb{P}^{N-1}}(n)$, and ρ_1 is a finite morphism. On the other hand,

$$\rho_1^* (\mathcal{O}_{\mathbb{P}^{N-1}}(n)) = \mathcal{O}_{Y_1}(nd).$$

Thus

$$H^0(\mathbb{P}^{N-1}, \mathcal{O}_{Y_1} \otimes \mathcal{O}_{\mathbb{P}^{N-1}}(n)) \cong H^0(Y_1, \mathcal{O}_{Y_1}(nd)) \cong R^G_{nd}$$

(where R^G_{nd} is the nd-graded piece of R^G). Thus the graded $K[x_1, \cdots, x_N]$-module associated to the coherent sheaf $\mathcal{O}_{Y_1}$ on $\mathbb{P}^{N-1}$ is $\oplus_{n \in \mathbb{Z}_+} R^G_{nd}$; and by the basic theorems of Serre, $\oplus_{n \in \mathbb{Z}_+} R^G_{nd}$ is of finite type over $K[x_1, \ldots, x_N]$. Now a dth power of any homogeneous element of R^G is in $\oplus_{n \in \mathbb{Z}_+} R^G_{nd}$, and thus R^G is integral over $K[x_1, \ldots, x_N]$. This implies ρ is finite, and the proof of Case 1 is complete.
Case 2. Let $f_1, \ldots, f_N$ be homogeneous of possibly different degrees, say $\deg f_i = d_i$. Set d equal to the least common multiple of $\{d_1, \ldots, d_N\}$ and $e_i = \frac{d}{d_i}$. Now let $g_i = f_i^{e_i}$; then $\{g_1, \ldots, g_N\}$ satisfies the hypothesis of Case 1; thus R^G is a finite $K[g_1, \ldots, g_N]$-module. Since

$$K[g_1, \ldots, g_N] \hookrightarrow K[f_1, \ldots, f_N] \hookrightarrow R^G,$$

we have that R^G is a finite $K[f_1, \ldots, f_N]$-module, and thus the map $\rho : \mathrm{Spec} R^G \to \mathrm{Spec} S$ has finite fibers.

□

Lemma 9.3.3. *Suppose $F : X \to Y$ is a surjective morphism of (irreducible) algebraic varieties, and U is an open subset of X such that*

1. *$F|_U : U \to Y$ is an immersion,*
2. $\dim U = \dim Y$.

Then F is birational.

Proof. The fact that $F|_U$ is an immersion implies that $F(U)$ is locally closed in Y, i.e., $F(U)$ is open in its closure. The fact that $\dim U = \dim Y$ implies that the closure of $F(U)$ is all of Y, i.e., $F(U)$ is open in Y. Thus U embeds as an open subset of Y. From this, the birationality of F follows. □

Combining the two previous lemmas, we can apply Zariski's Main Theorem to R^G as follows.

Theorem 9.3.4. *Let $X = Spec(R)$, $f_1, \ldots, f_N \in R^G$ be homogeneous, $S = K[f_1, \ldots, f_N]$, and $D = Spec(S)$. Let $\phi : X \to Spec(R^G)$ and $\psi : X \to \mathbb{A}^N$ be the map*

$$x \mapsto (f_1(x), \ldots, f_N(x)) .$$

Then D is isomorphic to $X /\!/ G$, with ψ as the canonical quotient map, if the following four conditions are satisfied.

1. *For every $x \in X^{ss}$, $\psi(x) \neq 0$.*
2. *There is a G-stable open subset U of X on which G operates freely with U/G as quotient, and ψ induces an immersion $U/G \to D$.*

3. $\dim D = \dim U/G$.
4. D is normal.

We now return to $X = M_{n,d}$ where $n > d$. Let $R = K[x_{ij}, 1 \leq i \leq n,\ 1 \leq j \leq d]$, the coordinate ring of X (i.e., $X = \mathrm{Spec}R$).

For $\underline{i} \in I_{d,n}$, let $f_{\underline{i}} \in R$ be defined such that

$$f_{\underline{i}}(A) = \det A_{\underline{i}},\ A \in X,$$

where $A_{\underline{i}}$ is the $d \times d$ minor of A with rows indexed by $i_1, \ldots, i_d$ (as in §9.2).

Let $G = SL_d(K)$. We have that G acts on X by right multiplication, and hence G acts on R. Recall that $f_{\underline{i}} \in R^G$, for all $\underline{i} \in I_{d,n}$.

Remark 9.3.5. Let $\pi : X \to \bigwedge^d(K^n)$, $\pi(A) = a_1 \wedge \ldots \wedge a_d$, where $a_1, \ldots, a_d$ are the columns of A (viewed here as elements of K^n). Recall from the section on integral schemes (the end of §5.3.2) that $\widehat{G_{d,n}}$ is identified with the image of π; and thus $K[G_{d,n}]$ is identified with the subring of $K[x_{ij}, 1 \leq i \leq n,\ 1 \leq j \leq d]$ generated by $\{f_{\underline{i}} \mid \underline{i} \in I_{d,n}\}$.

We now give the connection to $G_{d,n}$.

Theorem 9.3.6. *Let X and G be as above, then $X /\!\!/ G$ is isomorphic to $\widehat{G_{d,n}}$, the cone over the Grassmannian $G_{d,n}$.*

Proof. Let $N = \binom{n}{d}$ (which is the cardinality of $I_{d,n}$). We define $\psi : X \to \mathbb{A}^N$ such that

$$\psi(A) = \left(f_{\underline{i}}(A),\ \underline{i} \in I_{d,n}\right), \quad A \in X.$$

Let $D = \mathrm{Spec}(K[f_{\underline{i}},\ \underline{i} \in I_{d,n}])$; note that D is both the image of ψ as well as $\widehat{G_{d,n}}$ (as discussed in Remark 9.3.5). We verify the four hypotheses of Theorem 9.3.4 for ψ.

First, we note that if $x \in X^{ss}$, then $x \in X^o$ by Lemma 9.2.7, and thus there exists some $\underline{i} \in I_{d,n}$ such that $f_{\underline{i}}(x) \neq 0$, implying $\psi(x) \neq 0$. Thus hypothesis (1) of Theorem 9.3.4 holds.

To show that hypothesis (2) holds, we let $G' = GL_d(K)$. Set

$$U = \{A \in X \mid f_{(1,\ldots,d)}(A) \neq 0\}.$$

Clearly U is a G-stable open subset of X.

Claim. G operates freely on U, $U \to U/G$ is a G-principal fiber space, and ψ induces an immersion $U/G \to \mathbb{A}^N$.

To prove the claim, we first observe that we have a G-equivariant isomorphism

$$(*) \qquad U \cong G' \times M_{s,d}$$

where $s = n - d$. From this it is clear that G operates freely on U. Further, we see that U/G may be identified with the fiber space with base G'/G, and fiber $M_{s,d}$, associated to the principal fiber space $G' \to G'/G$. It remains to show that $\psi : U/G \to \mathbb{A}^N$ is an immersion. We shall show that both $\psi : U/G \to \mathbb{A}^N$ and its differential $d\psi$ are injective.

Let $y_1, y_2 \in U/G$ be such that $\psi(y_1) = \psi(y_2)$. Let $x_1, x_2 \in U$ be lifts for y_1, y_2, respectively. Using $(*)$, we may identify x_1, x_2 as

$$x_1 = \begin{bmatrix} A_1 \\ M_1 \end{bmatrix}, \quad x_2 = \begin{bmatrix} A_2 \\ M_2 \end{bmatrix}$$

where A_1, A_2 are $d \times d$ invertible matrices, and M_1, M_2 are $(n-d) \times d$ matrices. Since $\psi(y_1) = \psi(y_2)$, we have $\det A_1 = \det A_2$, and thus $g = (A_2)^{-1} A_1 \in G$. Hence in U/G, we may suppose that

$$y_1 = \begin{bmatrix} A_1 \\ M_1 \end{bmatrix}, \quad y_2 = \begin{bmatrix} A_1 \\ M_2 \end{bmatrix}$$

Let $\det(A_1) = c$, and let α be a dth root of c. Define $g' = \frac{1}{\alpha} A_1 \in G$; then

$$A_1 g'^{-1} = \alpha I = \begin{bmatrix} \alpha & 0 & \cdots & 0 \\ 0 & \alpha & \cdots & 0 \\ \vdots & \vdots & \ddots & \vdots \\ 0 & 0 & \cdots & \alpha \end{bmatrix}.$$

Thus we may suppose (in U/G), $y_1 = \begin{bmatrix} \alpha I \\ B_1 \end{bmatrix}$ and $y_2 = \begin{bmatrix} \alpha I \\ B_2 \end{bmatrix}$. Let k be such that $d < k \leq n$, and i be such that $1 \leq i \leq d$. Define $\tau_{ik} = (1, \ldots, i-1, i+1, \ldots, d, k)$. We have $f_{\tau_{ik}}(y_1) = f_{\tau_{ik}}(y_2)$ (since $\psi(y_1) = \psi(y_2)$); but on the other hand,

$$f_{\tau_{ik}}(y_1) = (-1)^{d-i} \alpha^{d-1} b_{ki}^{(1)}, \quad f_{\tau_{ik}}(y_2) = (-1)^{d-i} \alpha^{d-1} b_{ki}^{(2)}.$$

From this, it follows that $y_1 = y_2$ implying that $\psi : U/G \to \mathbb{A}^N$ is injective.

To see that $d\psi$ is injective, note that the above argument remains valid for the points over $K[\epsilon]$, the algebra of dual numbers ($= K \oplus K\epsilon$, the K-algebra with one generator ϵ, and one relation $\epsilon^2 = 0$). This completes the proof of the claim, and thus hypothesis (2) of Theorem 9.3.4 holds.

For hypothesis (3), we now show that $\dim D = \dim U/G$. Recall that $\dim D = \dim \widehat{G_{d,n}} = d(n-d) + 1$. On the other hand,

$$\dim U/G = \dim U - \dim G = nd - (d^2 - 1) = d(n-d) + 1.$$

Thus hypothesis (3) of Theorem 9.3.4 holds.

Finally, we have that D is normal by Theorem 6.3.9, and hypothesis (4) of Theorem 9.3.4 holds, and we conclude that $X \mathbin{/\!/} G = \widehat{G_{d,n}}$. □

Now that we have the identification $X \mathbin{/\!/} G = \widehat{G_{d,n}}$, we can use the results of Chapter 5 to conclude the "Fundamental" Theorems of Classical Invariant Theory (in any characteristic) (cf. [88]).

Theorem 9.3.7 (First Fundamental Theorem). *The ring of invariants R^G is generated by $\{f_{\underline{i}},\ \underline{i} \in I_{d,n}\}$.*

Theorem 9.3.8 (Second Fundamental Theorem). *The ideal of relations among the generators $\{f_{\underline{i}},\ \underline{i} \in I_{d,n}\}$ is generated by the Plücker quadratic relations.*

Corollary 9.3.9. *The ring of invariants R^G has a basis consisting of standard monomials in $\{f_{\underline{i}},\ \underline{i} \in I_{d,n}\}$.*

Chapter 10
Determinantal Varieties

In this chapter, we give an exposition of the classical determinantal varieties and bring out their relationship to Schubert varieties as well as certain GL_n-actions appearing in Classical Invariant Theory.

10.1 Determinantal Varieties

Let $Z = M_{m,n}(K)$, the space of all $m \times n$ matrices with entries in K. We shall identify Z with $\mathbb{A}^{mn}$. We have $K[Z] = K[x_{i,j},\ 1 \le i \le m,\ 1 \le j \le n]$.

10.1.1 The determinantal variety D_t

Let $X = (x_{ij}),\ 1 \le i \le m,\ 1 \le j \le n$ be an $m \times n$ matrix of indeterminates. Let $A \subset \{1, \cdots, m\},\ B \subset \{1, \cdots, n\},\ \#A = \#B = s$, where $s \le \min\{m, n\}$. We shall denote by $[A|B]$ the s-minor of X with row indices given by A, and column indices given by B. Fix $t,\ 1 \le t \le \min\{m, n\}$; let $I_t(X)$ denote the ideal in $K[x_{i,j}]$ generated by $\{[A|B],\ A \subset \{1, \cdots, m\},\ B \subset \{1, \cdots, n\},\ \#A = \#B = t\}$. Let $D_t(K)$ (or just D_t) be the *determinantal variety*, a closed subvariety of Z, with $I_t(X)$ as the defining ideal.

V. Lakshmibai, J. Brown, *The Grassmannian Variety*,
Developments in Mathematics 42, DOI 10.1007/978-1-4939-3082-1_10

10.1.2 Relationship between determinantal varieties and Schubert varieties

Let $G = SL_n(K)$. Let r, d be such that $r+d = n$. Let us identify (cf. Definition 6.4.1) the opposite cell O_d^- in $G/P_d(\cong G_{d,n})$ as

$$O_d^- = \left\{ \begin{pmatrix} I_d \\ X \end{pmatrix} \right\},$$

X being an $r \times d$ matrix. We have a bijection between

$$\{\text{Plücker co-ordinates } p_{\underline{i}},\ \underline{i} \neq \{1, 2, \cdots, d\}\} \text{ and } \{\text{minors of } X\}$$

(note that if $\underline{i} = \{1, 2, \cdots, d\}\}$, then $p_{\underline{i}} =$ the constant function 1).

Example 10.1.1. Take $r = 3 = d$. We have,

$$\mathcal{O}_3^- = \left\{ \begin{pmatrix} I_3 \\ X_{3\times 3} \end{pmatrix} \right\}.$$

We have $p_{(1,2,4)} = [\{1\}|\{3\}]$, $p_{(2,4,6)} = [\{1, 3\}|\{1, 3\}]$.

In the sequel, given a Plücker coordinate $p_{\underline{i}}$ (on $\mathcal{O}_d^-$), we shall denote the associated minor (of $X_{r,d}$) by $\Delta_{\underline{i}}$.

Let $Z_t = \{\underline{i} \in I_{d,n} \mid \Delta_{\underline{i}}$ is a t minor of $X\}$. The partial order on the Plücker coordinates induces a partial order on Z_t.

Lemma 10.1.2. *Let τ be the d-tuple, $\tau = (1, 2, \cdots, d-t, d+1, d+2, \cdots, d+t)$. Then $\tau \in Z_t$. Further, τ is the unique smallest element in Z_t.*

Proof. Clearly $\tau \in Z_t$. Let Δ be a t-minor of X, and $p_{\underline{i}}$ the associated Plücker coordinate. Let $\underline{i} = (i_1, \cdots, i_d)$. We have, for $1 \leq k \leq d-t$, $i_k \leq d$, and for $d-t+1 \leq k \leq d$, $i_k > d$. Clearly τ is the smallest such d-tuple. □

Remark 10.1.3. With τ as in Lemma 10.1.2, note that the associated minor Δ_τ of X has row and column indices given by $\{d+1, d+2, \cdots, d+t\}$, $\{d+1-t, d+2-t, \cdots, d\}$, respectively, i.e., the right most top corner t minor of X, the rows of X being indexed as $d+1, \ldots, n$.

Remark 10.1.4. With τ as in Lemma 10.1.2, note that τ is the smallest d-tuple $(j_1, \cdots, j_d)$ such that $j_{d-t+1} \geq d+1$.

Lemma 10.1.5. *Let τ be as in Lemma 10.1.2. Let*

$$N_t = \{\underline{i} \in I_{d,n} \mid p_\tau\,|_{X_{\underline{i}}} = 0\}.$$

Let ϕ be the d-tuple, $\phi = (t, t+1, \cdots, d, n+2-t, n+3-t, \cdots, n)$ (note that ϕ consists of the two blocks $[t, d]$, $[n+2-t, n]$ of consecutive integers - here, for $i < j$, $[i, j]$ denotes the set $\{i, i+1, \cdots, j\}$). Then ϕ is the unique largest element in N_t.

Proof. Let $\underline{i} \in N_t$, say $\underline{i} = (i_1, \cdots, i_d)$. We have, $p_\tau\,|_{X_{\underline{i}}} = 0$ if and only if $\underline{i} \not\geq \tau$, i.e., if and only if $i_{d-t+1} \not\geq d+1$ (cf. Remark 10.1.4), i.e., $i_{d-t+1} \leq d$. Now it is easily checked that ϕ is the largest d-tuple $(j_1, \cdots, j_d)$ such that $j_{d-t+1} \leq d$. □

Remark 10.1.6. Let ϕ be as in Lemma 10.1.5. As observed in the proof of Lemma 10.1.5, we have ϕ is the largest d-tuple $(j_1, \cdots, j_d)$ such that $j_{d-t+1} \leq d$.

Corollary 10.1.7. *Let $\underline{i} \in Z_t$. Then $p_{\underline{i}}\,|_{X_\phi} = 0$, ϕ being as in Lemma 10.1.5.*

Proof. We have (cf. Lemma 10.1.5), $p_\tau\,|_{X_\phi} = 0$, τ being as in Lemma 10.1.2 and hence $\phi \not\geq \tau$ which in turn implies (in view of Lemma 10.1.2), $\phi \not\geq \underline{i}$. Hence we obtain $p_{\underline{i}}\,|_{X_\phi} = 0$. □

Theorem 10.1.8. *Let ϕ be as in Lemma 10.1.5, and let $Y_\phi = O_d^- \cap X_\phi$ (the opposite cell in X_ϕ). Then $D_t \cong Y_\phi$.*

Proof. Let I_ϕ be the ideal defining Y_ϕ in O_d^-. We have that I_ϕ is generated by $M_t := \{p_{\underline{i}},\ \underline{i} \not\leq \phi\}$. Also, $I(D_t)$ (the ideal defining D_t in O_d^-) is generated by $\{p_{\underline{i}},\ \underline{i} \in Z_t\}$.

Let $\underline{i} \in Z_t$, say, $\underline{i} = (i_1, \cdots, i_d)$. We have (cf. Lemma 10.1.5) $\underline{i} \not\leq \phi$, and hence $p_{\underline{i}} \in M_t$.

Let now $p_{\underline{i}} \in M_t$, say, $\underline{i} = (i_1, \cdots, i_d)$. This implies that $i_{d-t+1} \geq d+1$ (cf. Remark 10.1.6), and hence it corresponds to an s-minor in X, where $s \geq t$. From this it follows that $p_{\underline{i}}$ is in the ideal generated by $\{p_{\underline{i}},\ \underline{i} \in Z_t\}$, i.e., $p_{\underline{i}}$ belongs to $I(D_t)$.

Thus we have shown $I_\phi = I(D_t)$ and the result follows from this. □

Corollary 10.1.9. *The determinantal variety D_t is normal, Cohen–Macaulay of dimension $(t-1)(n-(t-1))$, where note that $n = r + d$.*

Proof. The normality and Cohen–Macaulayness of Schubert varieties (cf. Chapter 6) imply corresponding properties for D_t (in view of Theorem 10.1.8). Regarding $\dim D_t$, we have $\dim D_t = \dim X_\phi = (t-1)(n-(t-1))$ (recall from §5.3 that if $w = (a_1, \cdots, a_d) \in I_{d,n}$, then $\dim X(w) = \sum_{i=1}^{i=d} (a_i - i)$). □

10.2 Standard Monomial Basis for $K[D_t]$

In this section, we describe a standard monomial basis for $K[D_t]$, using Theorem 10.1.8. To begin with, we set up a bijection between $I_{d,n} \setminus \{(1, 2, \ldots, d)\}$ and $\{$all minors of $X\}$ as follows:

Let $n = r + d$. Given $\underline{i} \in I_{d,n}$, let m be such that $i_m \leq d$, $i_{m+1} > d$. Set

$$A_{\underline{i}} = \{n+1-i_d, n+1-i_{d-1}, \ldots, n+1-i_{m+1}\},$$

$$B_{\underline{i}} = \text{ the complement of } \{i_1, i_2, \ldots, i_m\} \text{ in } \{1, 2, \ldots, d\}.$$

Define $\theta : I_{d,n} \setminus \{(1, 2, \ldots, d)\} \to \{\text{all minors of } X\}$ by setting $\theta(\underline{i}) = [A_{\underline{i}}|B_{\underline{i}}]$. Clearly, θ is a bijection.

10.2.1 The partial order $\succeq$

Define a partial order on the set of all minors of X as follows:

Let $[A|B]$, $[A'|B']$ be two minors of size s, s', respectively. Let $A = \{a_1, \cdots, a_s\}$, $B = \{b_1, \cdots, b_s\}$, $A' = \{a'_1, \cdots, a'_{s'}\}$, $B = \{b'_1, \cdots, b'_{s'}\}$. We define $[A|B] \succeq [A'|B']$ if $s \leq s'$, $a_j \geq a'_j$, $b_j \geq b'_j$, $1 \leq j \leq s$. Note that the bijection θ reverses the respective partial orders, i.e., given $\underline{i}, \underline{i}' \in I_{d,n}$, we have $\underline{i} \leq \underline{i}'$ if and only if $\theta(\underline{i}) \succeq \theta(\underline{i}')$. Using this partial order, we define *standard monomials* in $[A, B]$'s.

Definition 10.2.1. A monomial $[A_1, B_1] \cdots [A_s, B_s], s \in \mathbb{N}$ is standard if $[A_1, B_1] \succeq \cdots \succeq [A_s, B_s]$.

In view of §5.4 and Theorem 10.1.8, we obtain the following theorem.

Theorem 10.2.2. *Standard monomials in $[A, B]$'s with $\#A \leq t - 1$ form a basis for $K[D_t]$, the algebra of regular functions on D_t.*

Proof. Theorem 10.1.8 identifies D_t as the principal open set $X(\phi)_{p_{id}}$; hence we get an identification

$$K[D_t] \cong K[X(\phi)]_{(p_{id})}$$

where $K[X(\phi)]_{(p_{id})}$ denotes the homogeneous localization; note that p_{id} is a lowest weight vector in $H^0(G_{d,n}, L(\omega_d))$. The fact that standard monomials $\{p_{\tau_1} \cdots p_{\tau_m}, m \in \mathbb{N}, \phi \geq \tau_1\}$ generate $K[X(\phi)]$ implies that

$$\left\{ \frac{p(\tau_1)}{p_{id}} \cdots \frac{p(\tau_m)}{p_{id}}, m \in \mathbb{N}, \phi \geq \tau_1 \right\}$$

generates $K[D_t]$. Note that the condition that $\phi \geq \tau$ corresponds to the condition that the minor corresponding to τ has size $\leq t - 1$ (note that under the order-reversing bijection θ, ϕ corresponds to the pair $((1, \cdots, t-1), (1, \cdots, t-1))$). The linear independence of standard monomials $\{p_{\tau_1} \cdots p_{\tau_m}, m \in \mathbb{N}, \phi \geq \tau_1\}$ clearly implies the linear independence of $\{\frac{p(\tau_1)}{p_{id}} \cdots \frac{p(\tau_m)}{p_{id}}, m \in \mathbb{N}, \phi \geq \tau_1\}$. Further, clearly $p_{\tau_1} \cdots p_{\tau_m}$ is standard if and only if $\frac{p(\tau_1)}{p_{id}} \cdots \frac{p(\tau_m)}{p_{id}}$ is. The result now follows from this. □

In particular, taking $t = 1 + min\{m, n\}$, we recover the result due to Doubilet-Rota-Stein [19, Theorem 2].

Theorem 10.2.3. *Standard monomials in $[A, B]$'s form a basis for $K[Z]$ ($\cong K[x_{ij}, 1 \leq i \leq m, 1 \leq j \leq n]$).*

10.2.2 Cogeneration of an Ideal

As above, let X be a matrix of size $r \times d$, and let $\geq$ be a partial order on the set of all minors of X.

Definition 10.2.4. Given a minor M of X, the ideal I_M of $K[x_{ij}, 1 \leq i \leq r, 1 \leq j \leq d]$ generated by $\{M' \mid M' \text{ a minor of } X \text{ such that } M' \not\geq M\}$ is called the ideal *cogenerated* by M (for the given partial order).

Proposition 10.2.5. *Given a minor M of X, let τ_M be the element of $I_{d,n}$ such that $\theta(\tau_M) = M$. Then $I_M = I(Y_{\tau_M})$, where $Y_{\tau_M} = O_d^- \cap X_{P_d}(\tau_M)$ and $I(Y_{\tau_M})$ is the ideal defining Y_{τ_M} as a closed subvariety of O_d^-.*

Proof. We have (cf. Theorem 10.1.8), $I(Y_{\tau_M})$ is generated by $\{p_{\underline{i}} \mid \underline{i} \in I_{d,n}, \tau_M \not\geq \underline{i}\}$. The required result follows from this in view of the bijection θ. □

Corollary 10.2.6. *The set $\{I_M, M \text{ a minor of } X\}$ is precisely the set of ideals of the opposite cells in Schubert varieties in $G_{d,n}$.*

Proof. The result follows in view of Proposition 10.2.5 and the bijection θ between $I_{d,n} \setminus \{(1, 2, \ldots, d)\}$ and the minors of X. □

Corollary 10.2.7. *Let M be the $(t-1)$-minor of X consisting of the first $(t-1)$-rows of X and the first $(t-1)$-columns of X. Then $I_t(X)$ (the ideal of D_t) equals I_M.*

Proof. We have (cf. Theorem 10.1.8), $I_t(X) = I(Y_\phi)$, ϕ being as in Theorem 10.1.8. Now, under the bijection θ, we have $\theta(\phi) = [\{1, 2, \cdots, t-1\} \mid \{1, 2, \cdots, t-1\}]$. The result now follows from Proposition 10.2.5. □

10.3 Gröbner Bases for Determinantal Varieties

In this section, we determine a Gröbner Basis for the ideal of a determinantal variety.

First, we introduce a total order on the variables as follows:

$$x_{m1} > x_{m2} > \cdots > x_{mn} > x_{m-1\,1} > x_{m-1\,2} > \ldots$$
$$\cdots > x_{m-1\,n} > \cdots > x_{11} > x_{12} > \cdots > x_{1n}.$$

This induces a total order, namely the lexicographic order, on the set of monomials in $K[X] = K[x_{11}, \ldots, x_{mn}]$, denoted by $\prec$. Throughout the rest of this section, for a polynomial f, $\mathrm{in}(f)$ will denote the initial term of f with respect to $\prec$ (where initial term is the largest monomial). Note that the initial term (with respect to $\prec$) of a minor of X is equal to the product of its elements on the skew diagonal. Similarly, given an ideal $I \subset K[X]$, we denote by $\mathrm{in}(I)$ the ideal generated by the initial terms of the elements in I with respect to $\prec$.

We recall the following from [29].

Theorem 10.3.1. *Let $M = [i_1, \ldots, i_r | j_1, \ldots, j_r]$ be a minor of X, and I the ideal of $K[X]$ cogenerated by M. For $1 \leq s \leq r+1$, let G_s be the set of all s-minors $[i'_1, \ldots, i'_s | j'_1, \ldots, j'_s]$ satisfying the conditions*

$$i'_s \leq i_r, i'_{s-1} \leq i_{r-1}, \ldots, i'_2 \leq i_{r-s+2}, \tag{1}$$

$$j'_{s-1} \geq j_{s-1}, \ldots, j'_2 \geq j_2, j'_1 \geq j_1$$

$$\text{if } s \leq r, \text{ then } i'_1 > i_{r-s+1} \text{ or } j'_s < j_s. \tag{2}$$

Then the set $G = \bigcup_{i=1}^{r+1} G_i$ is a Gröbner basis for the ideal I with respect to the monomial order $\prec$.

Theorem 10.3.2. *Let $t \in \mathbb{Z}$, $1 \leq t \leq \min\{m, n\}$. Let*

$$\mathcal{F}_t = \{M' \mid M' \text{ is a minor of } X \text{ of size } t\}.$$

Then $\mathcal{F}_t$ is a Gröbner basis for $I_t(X)$.

Proof. We have D_t is generated by $\mathcal{F}_t$. Let M be the $(t-1)$-minor $[A|B]$ of X, where $A = \{m+2-t, m+3-t, \cdots, m\}$, $B = \{1, 2, \cdots, t-1\}$. Let us write $A = \{i_1, \cdots, i_r\}$, $B = \{j_1, \cdots, j_r\}$.We have $r = t-1$, $i_l = m+l-r$, $j_l = l$, $1 \leq l \leq r$.

Let $M' \in G$, G being as in Theorem 10.3.1, say, $M' = [i'_1, \ldots, i'_s | j'_1, \ldots, j'_s]$ be a minor of X of size s, $s \leq r+1 (= t)$.The inequalities regarding i's and i''s in condition (1) of Theorem 10.3.1 are redundant, since $i_l = m+l-r$, $1 \leq l \leq r$. Similarly, the inequalities regarding j's and j''s in condition (1) of Theorem 10.3.1 are again redundant (since $j_l = l$, $1 \leq l \leq r$); also, condition (2) reduces to the condition that if $s \leq r(= t-1)$, then $i'_1 > i_{r-s+1}$ (since $j_s = s$). Therefore, for the above choice of M, the conditions (1) and (2) of Theorem 10.3.1 are equivalent to

$$\text{if } s \leq r\, (= t-1), \text{ then } i'_1 > i_{r-s+1}\, (= i_{t-s}).$$

Let $s \leq t-1$. The condition that $i'_1 > i_{t-s}\, (= m+1-s)$ implies $i'_s > m+1$, which is not possible. Hence $G_s = \emptyset$, if $s \leq t-1$; thus we obtain $G = G_t$. Now for $s = t$, as discussed above, condition (1) of Theorem 10.3.1 is satisfied by any t-minor M' while condition (2) is vacuous. Hence we obtain that G_t is the set of all t-minors, i.e., $G = \mathcal{F}_t$. The result now follows from Theorem 10.3.1. □

We conclude this section with the description of the singular locus of the determinantal variety.

Theorem 10.3.3. *The Singular Locus of D_t is D_{t-1}.*

Proof. Let $\phi = (t, t+1, \ldots, d, n+2-t, \ldots, n)$; from Theorem 10.1.8 we have $D_t \cong Y_\phi$. Let us use the formulation of SingY_ϕ found in Theorem 6.5.9 and Example 6.5.10. As an element of $I_{d,n}$, ϕ corresponds to the partition $\mathbf{a} = \left((n-d)^{t-1}, (t-1)^{d+t-t}\right)$. The partition $\mathbf{a}$ has only one "hook," and thus

$\mathrm{Sing}Y_{\mathbf{a}} = Y_{\mathbf{b}}$, where $\mathbf{b} = \left((n-d)^{t-2}, (t-2)^{d+2-t}\right)$ (we note that the results on the singular locus of the Schubert variety $X_{\mathbf{a}}$ apply to the opposite cell $Y_{\mathbf{a}}$).

The partition $\mathbf{b}$ correspond to $\tau \in I_{d,n}$, $\tau = (t-1, t, \ldots, d, n+3-t, \ldots, n)$, and $Y_\tau \cong D_{t-1}$ by Theorem 10.1.8. The result follows. □

10.4 Connections with Classical Invariant Theory

In Chapter 9, we examined the results of $SL_d(K)$-actions appearing in classical invariant theory and their connections to the Grassmannian. In this section, we consider $GL_n(K)$-actions appearing in classical invariant theory (cf. De Concini-Procesi [17], Weyl [88]), and we will see a connection to the determinantal variety.

Let $V = K^n$, and

$$X = \underbrace{V \oplus \cdots \oplus V}_{m \text{ copies}} \oplus \underbrace{V^* \oplus \cdots \oplus V^*}_{q \text{ copies}}.$$

The $GL(V)$-action on X: Writing $\underline{u} = (u_1, u_2, \ldots, u_m)$ with $u_i \in V$ and $\underline{\xi} = (\xi_1, \xi_2, \ldots, \xi_q)$ with $\xi_i \in V^*$, we shall denote the elements of X by $(\underline{u}, \underline{\xi})$. The (natural) action of $GL(V)$ on V induces an action of $GL(V)$ on V^*, namely, for $\xi \in V^*, g \in GL(V)$, denoting $g \cdot \xi$ by ξ^g, we have

$$\xi^g(v) = \xi(g^{-1}v), v \in V$$

The diagonal action of $GL(V)$ on X is given by

$$g \cdot (\underline{u}, \underline{\xi}) = (g\underline{u}, \underline{\xi}^g) = (gu_1, gu_2, \ldots, gu_m, \xi_1^g, \xi_2^g, \ldots, \xi_q^g),\ g \in G, (\underline{u}, \underline{\xi}) \in X.$$

The induced action on $K[X]$ is given by

$$(g \cdot f)(\underline{u}, \underline{\xi}) = f(g^{-1}(\underline{u}, \underline{\xi})), f \in K[X],\ g \in GL(V).$$

Consider the functions $\varphi_{ij} : X \longrightarrow K$ defined by $\varphi_{ij}((\underline{u}, \underline{\xi})) = \xi_j(u_i)$, $1 \leq i \leq m$, $1 \leq j \leq q$. Each φ_{ij} is a $GL(V)$-invariant: For $g \in GL(V)$, we have,

$$\begin{aligned}
(g \cdot \varphi_{ij})((\underline{u}, \underline{\xi})) &= \varphi_{ij}(g^{-1}(\underline{u}, \underline{\xi})) \\
&= \varphi_{ij}((g^{-1}u, \xi^{g^{-1}}) \\
&= \xi_j^{g^{-1}}(g^{-1}u_i) \\
&= \xi_j(u_i) \\
&= \varphi_{ij}((\underline{u}, \underline{\xi})).
\end{aligned}$$

It is convenient to have a description of the above action in terms of coordinates. So with respect to a fixed basis, we write the elements of V as row vectors and those of V^* as column vectors. Thus denoting by $M_{a,b}$ the space of $a \times b$ matrices with entries in K, X can be identified with the affine space $M_{m,n} \times M_{n,q}$. The action of $GL_n(K)$ $(= GL(V))$ on X is then given by

$$A \cdot (U, W) = (UA, A^{-1}W),\ A \in GL_n(K),\ U \in M_{m,n},\ W \in M_{n,q}.$$

And the action of $GL_n(K)$ on $K[X]$ is given by

$$(A \cdot f)(U, W) = f\left(A^{-1}(U, W)\right) = f\left(UA^{-1}, AW\right), f \in K[X].$$

Writing $U = \left(u_{ij}\right)$ and $W = (\xi_{kl})$ we denote the coordinate functions on X by u_{ij} and ξ_{kl}. Further, if u_i denotes the i-th row of U and ξ_j the j-th column of W, the invariants φ_{ij} described above are nothing but the entries $\langle u_i, \xi_j\rangle$ $(= \xi_j(u_i))$ of the product UW. In the sequel, we shall denote $\varphi_{ij}(\underline{u}, \underline{\xi})$ also by $\langle u_i, \xi_j\rangle$.

The Function $p(A, B)$: For $A \in I(r, m), B \in I(r, q), 1 \leq r \leq n$, let $p(A, B)$ be the regular function on X: $p(A, B)((\underline{u}, \underline{\xi}))$ is the determinant of the $r \times r$-matrix $(\langle u_i, \xi_j\rangle)_{i\in A, j\in B}$. Let S be the subalgebra of R^G generated by $\{p(A, B)\}$. We shall now show (using Theorem 9.3.4) that S is in fact equal to R^G.

10.4.1 The First and Second Fundamental Theorems of Classical Invariant Theory (cf. [88]) for the action of $GL_n(K)$

Theorem 10.4.1. *Let $G = GL_n(K)$. Let X be as above. The morphism ψ : $X \to M_{m,q}$, $(\underline{u}, \underline{\xi}) \mapsto \left(\varphi_{ij}(\underline{u}, \underline{\xi})\right)$ $(= (\langle u_i, \xi_j\rangle))$ maps X into the determinantal variety $D_{n+1}(M_{m,q})$, and the induced homomorphism ψ^* : $K[D_{n+1}(M_{m,q})] \to K[X]$ between the coordinate rings induces an isomorphism*

$$\psi^* : K[D_{n+1}(M_{m,q})] \to K[X]^G,$$

i.e., the determinantal variety $D_{n+1}(M_{m,q})$ is the categorical quotient of X by G.

Proof. Clearly, $\psi(X) \subseteq D_{n+1}(M_{m,q})$ (since $\psi(X) = Spec\, S$, and clearly $Spec\, S \subseteq D_{n+1}(M_{m,q})$ (since any $n + 1$ vectors in V are linearly independent)). We shall prove the result using Theorem 9.3.4. To be very precise, we shall first check the conditions (i)–(iii) of Theorem 9.3.4 for $\psi : X \to M_{m,q}$, deduce that the inclusion $Spec\, S \subseteq D_{n+1}(M_{m,q})$ is in fact an equality, and hence conclude the normality of $Spec\, S$ (condition (iv) of Theorem 9.3.4).

(i) Let $x = (\underline{u}, \underline{\xi}) = (u_1, \ldots, u_m, \xi_1, \ldots, \xi_q) \in X^{ss}$. Let $W_{\underline{u}}$ be the subspace of V spanned by x_i's and $W_{\underline{\xi}}$ the subspace of V^* spanned by ξ_j's. Assume if possible that $\psi((\underline{u}, \underline{\xi})) = 0$, i.e., $\langle u_i, \xi_j \rangle = 0$ for all i, j.

Case (a): $W_{\underline{\xi}} = 0$, i.e., $\xi_j = 0$ for all j.
Consider the one parameter subgroup $\Gamma = \{g_t, t \neq 0\}$ of $GL(V)$, where $g_t = tI_n$, I_n being the $n \times n$ identity matrix. Then $g_t \cdot x = g_t \cdot (\underline{u}, \underline{0}) = (t\underline{u}, \underline{0})$, so that $g_t \cdot x \to (0)$ as $t \to 0$. Thus the origin 0 is in the closure of $G \cdot x$, and consequently x is not semi-stable, which is a contradiction.
Case (b): $W_{\underline{\xi}} \neq 0$.
Since the case $W_{\underline{u}} = 0$ is similar to Case (a), we may assume that $W_{\underline{u}} \neq 0$. Also the fact that $W_{\underline{\xi}} \neq 0$ together with the assumption that $\langle x_i, \xi_j \rangle = 0$ for all i, j implies that $\dim W_{\underline{u}} < n$. Let $r = dim\, W_{\underline{u}}$ so that we have $0 < r < n$. Hence, we can choose a basis $\{e_1, \ldots, e_n\}$ of V such that $W_{\underline{u}} = \langle e_1, \ldots, e_r \rangle$, $r < n$, and $W_{\underline{\xi}} \subset \langle e^*_{r+1}, \ldots, e^*_n \rangle$, where $\{e^*_1, \ldots, e^*_n\}$ is the dual basis in V^*. Consider the one parameter subgroup $\Gamma = \{g_t, t \neq 0\}$ of $GL(V)$, where

$$g_t = \begin{pmatrix} tI_r & 0 \\ 0 & t^{-1}I_{n-r} \end{pmatrix}.$$

We have $g_t \cdot (\underline{u}, \underline{\xi}) = (t\underline{u}, t\underline{\xi}) \to 0$ as $t \to 0$. Thus, by the same reasoning as in Case (a), the point $(\underline{u}, \underline{\xi})$ is not semi-stable, which leads to a contradiction. Hence we obtain $\psi((\underline{u}, \underline{\xi})) \neq 0$. Thus (i) of Theorem 9.3.4 holds.

(ii) Let

$$U = \{(\underline{u}, \underline{\xi}) \in X \mid \{u_1, \ldots, u_n\}, \{\xi_1, \ldots, \xi_n\} \text{ are linearly independent}\}.$$

Clearly, U is a G-stable open subset of X.

Claim: G operates freely on U, $U \to U\, mod\, G$ is a G-principal fiber space, and ψ induces an immersion $U/G \to M_{m,q}$.

Proof of Claim: We have a G-equivariant identification

$$(\dagger) \qquad U \cong G \times G \times \underbrace{V \times \cdots \times V}_{(m-n) \text{ copies}} \times \underbrace{V^* \times \cdots \times V^*}_{(q-n) \text{ copies}}$$

from which it is clear that and G operates freely on U. Further, we see that $U\, mod\, G$ may be identified with the fiber space with base $G \times G\, mod\, G$ (G acting on $G \times G$ as $g \cdot (g_1, g_2) = (g_1 g, g^{-1} g_2), g, g_1, g_2 \in G$), and fiber $\underbrace{V \times \cdots \times V}_{(m-n) \text{ copies}} \times \underbrace{V^* \times \cdots \times V^*}_{(q-n) \text{ copies}}$ associated to the principal fiber space $G \times G \to (G \times G) / G$.

It remains to show that ψ induces an immersion $U/G \to \mathbb{A}^N$, i.e., to show that the map $\psi : U/G \to M_{m,q}$ and its differential $d\psi$ are both injective. We first prove that $\psi : U/G \to M_{m,q}$ is injective. Let x, x' in U/G be such that $\psi(x) = \psi(x')$.

Let $\eta, \eta' \in U$ be lifts for x, x', respectively. Using the identification (†) above, we may write

$$\eta = (A, u_{n+1}, \cdots, u_m, B, \xi_{n+1}, \cdots, \xi_q),\ A, B \in G,$$
$$\eta' = (A', u'_{n+1}, \cdots, u'_m, B', \xi'_{n+1}, \cdots, \xi'_q),\ A', B' \in G$$

(here, $u_i, 1 \leq i \leq n$ are given by the rows of A, while $\xi_i, 1 \leq i \leq n$ are given by the columns of B; similar remarks on u'_i, ξ'_i). The hypothesis that $\psi(x) = \psi(x')$ implies in particular that

$$\langle u_i, \xi_j \rangle = \langle u'_i, \xi'_j \rangle\ , 1 \leq i, j \leq n$$

which may be written as

$$AB = A'B'$$

This implies that $A' = A \cdot g$, where $g = BB'^{-1}$. Hence on U/G, we may suppose that

$$x = (u_1, \cdots, u_n, u_{n+1}, \cdots, u_m, \xi_1, \cdots, \xi_q),$$
$$x' = (u_1, \cdots, u_n, u'_{n+1}, \cdots, u'_m, \xi'_1, \cdots, \xi'_q)$$

where $\{u_1, \ldots, u_n\}$ is linearly independent.

For a given j, we have,

$$\langle u_i, \xi_j \rangle = \langle u_i, \xi'_j \rangle\ , 1 \leq i \leq n, \text{ implies } \xi_j = \xi'_j$$

(since $\{u_1, \ldots, u_n\}$ is linearly independent). Thus we obtain

$$(*) \qquad \xi_j = \xi'_j, \text{ for all } j.$$

On the other hand, we have (by definition of U) that $\{\xi_1, \ldots, \xi_n\}$ is linearly independent. Hence, fixing an i, $n + 1 \leq i \leq m$, we get

$$\langle u_i, \xi_j \rangle = \langle u'_i, \xi_j \rangle (= \langle u'_i, \xi'_j \rangle)\ , 1 \leq j \leq n, \text{ implies } u_i = u'_i.$$

Thus we obtain

$$(**) \qquad u_i = u'_i, \text{ for all } i.$$

The injectivity of $\psi : U/G \to M_{m,q}$ follows from $(*)$ and $(**)$.

To prove that the differential $d\psi$ is injective, we merely note that the above argument remains valid for the points over $K[\varepsilon]$, the algebra of dual numbers

($= K \oplus K\varepsilon$, the K-algebra with one generator ε, and one relation $\varepsilon^2 = 0$), i.e., it remains valid if we replace K by $K[\varepsilon]$, or in fact by any K-algebra. Thus (ii) of Theorem 9.3.4 holds.

(iii) We have

$$\dim U/G = \dim U - \dim G = (m+q)n - n^2 = \dim D_{n+1}(M_{m,q}).$$

The immersion $U/G \hookrightarrow \mathit{Spec}\, S (\subseteq D_{n+1}(M_{m,q}))$ together with the fact above that $\dim U/G = \dim D_{n+1}(M_{m,q})$ implies that $\mathit{Spec}\, S$ in fact equals $D_{n+1}(M_{m,q})$. Thus (iii) of Theorem 9.3.4 holds.

(iv) The normality of $\mathit{Spec}\, S (= D_{n+1}(M_{m,q}))$ follows from Corollary 10.1.9.

□

Combining the above theorem with Theorem 10.2.2, we obtain the following corollary.

Corollary 10.4.2. *Let X and G be as above. Let φ_{ij} denote the regular function $(\underline{u}, \underline{\xi}) \mapsto \langle u_i, \xi_j \rangle$ on X, $1 \leq i \leq m$, $1 \leq j \leq q$, and let f denote the $m \times q$ matrix (φ_{ij}). The ring of invariants $K[X]^G$ has a basis consisting of standard monomials in the regular functions $p_{\lambda,\mu}(f)$ with $\#\lambda \leq n$, where $\#\lambda = t$ is the number of elements in the sequence $\lambda = (\lambda_1, \ldots, \lambda_t)$ and $p_{\lambda,\mu}(f)$ is the t-minor with row indices $\lambda_1, \ldots, \lambda_t$ and column indices $\mu_1, \ldots, \mu_t$.*

As a consequence of the above theorem and corollary, we obtain the First and Second Fundamental Theorems of classical invariant theory (cf. [88]). Let notation be as above.

Theorem 10.4.3.

1. ***First Fundamental Theorem***
The ring of invariants $K[X]^{GL(V)}$ is generated by φ_{ij}, $1 \leq i \leq m, 1 \leq j \leq q$.
2. ***Second Fundamental Theorem***
The ideal of relations among the generators in (1) is generated by the $(n+1)$-minors of the $m \times q$-matrix (φ_{ij}).

Further we have a standard monomial basis for the ring of invariants, as follows.

Theorem 10.4.4. *The ring of invariants $K[X]^{GL(V)}$ has a basis consisting of standard monomials in the regular functions $p(A, B)$, $A \in I(r, m)$, $B \in I(r, q)$, $r \leq n$.*

Chapter 11
Related Topics

In this chapter, we give a brief account of some of the topics that are related to flag and Grassmannian varieties.

11.1 Standard Monomial Theory for a General G/Q

We saw in Chapter 5 that the Grassmannian variety is a G/P, $G = SL(n)$, and P a suitable maximal parabolic subgroup. Recall that the classical result due to Hodge for the Grassmannian consists of constructing bases for $H^0(X, L^m), m \in \mathbb{Z}_+$ (X being the Grassmannian, and $L(= \mathcal{O}_X(1))$ being the ample generator of $\operatorname{Pic} X(\cong \mathbb{Z})$, the Picard group of X), in terms of "standard monomials" in the Plücker coordinates; this result was extended to the flag variety $Fl_n(\cong SL(n)/B)$, in Chapter 8, by developing a standard monomial theory for the flag variety. Generalization of Hodge's result to any G/Q, G semisimple and Q a parabolic subgroup of G (namely, construction of explicit bases for $H^0(G/Q, L)$ for all ample line bundles on G/Q), has been carried out in [56, 67] by developing a standard monomial theory for G/Q; in fact, a standard monomial basis for $H^0(X, L)$, X a Schubert variety, has also been constructed.

The standard monomial theory has led to very many interesting geometric and representation-theoretic consequences: normality, Cohen–Macaulayness properties for Schubert varieties, determination of singular loci of Schubert varieties, generalization of Littlewood-Richardson rule, branching rule, etc. See [4, 43, 44, 54, 57, 63–65] for details. For results on singularities of Schubert varieties, see also [12, 41, 62, 91].

V. Lakshmibai, J. Brown, *The Grassmannian Variety*,
Developments in Mathematics 42, DOI 10.1007/978-1-4939-3082-1_11

11.2 The Cohomology and Homology of the Flag Variety

We state below the main results concerning the cohomology and homology of the flag variety $Fl(n)$. For details, the reader may refer to [23]. Let X be an algebraic variety. Let H_iX and H^iX be the ith singular homology and cohomology of X with integer coefficients (see [83] for the definition of H_iX, H^iX, X being a topological space). The cohomology $H^*X = \oplus H^iX$ has a ring structure, $H^iX \otimes H^jX \to H^{i+j}X$, $\alpha \otimes \beta \mapsto \alpha \cup \beta$ (the cup product); it is an associative skew-commutative ring with $1 \in H^0X$. The homology $H_*X = \oplus H_iX$ is a left module over H^*X defined by $H^iX \otimes H_jX \to H_{j-i}X$, $\alpha \otimes \beta \mapsto \alpha \cap \beta$ (the cap product).

11.2.1 A $\mathbb{Z}$-basis for $H^*(Fl(n))$

Now take $X = Fl(n)$, the set of complete flags:

$$\{F_\bullet : (0) = V_0 \subset V_1 \subset \cdots \subset V_n = \mathbb{C}^n, dim\, V_i = i\}.$$

Let E_X be the trivial bundle of rank n over X. There is a universal filtration of submodules $(0) = U_0 \subset U_1 \subset \cdots \subset U_n = E_X = E$, say, $rank\, U_i = i$. Over a point in X (corresponding to a flag $F_\bullet$) the fibers of these bundles are simply the vector spaces occurring in $F_\bullet$. Let $L_i = U_i/U_{i-1}$, $x_i = c_1(L_i)$ (the Chern class of L_i, for reference see [28, App. A]).

Theorem 11.2.1. *The set* $\{x_1^{i_1}x_2^{i_2}\ldots x_n^{i_n}, 0 \le i_j \le n-j\}$ *is a* $\mathbb{Z}$*-basis for* $H^*(X)$.

11.2.2 A presentation for the $\mathbb{Z}$-algebra $H^*(Fl(n))$

Let $X = Fl(n)$ and $E = E_X$, the trivial bundle of rank n over X. From above, we have that E has a filtration with line bundles L_i as quotients. We have that $c_i(E)$ (the ith Chern class of E, see [28, App. A]) is the ith elementary symmetric polynomial in $c_1(L_i)(= x_i)$, $1 \le i \le n$. But $c_i(E) = 0$ for all i, since E is the trivial bundle. Thus we obtain $e_i(x_1,\ldots,x_n) = 0$, where $e_i(x_1,\ldots,x_n) = \sum_{i_1<\cdots<i_j} x_{i_1}\ldots x_{i_j}$, the ith elementary symmetric polynomial.

Theorem 11.2.2. *The* $\mathbb{Z}$*-algebra* $H^*(Fl(n))$ *has a set of algebra generators* $\{x_i, 1 \le i \le n\}$ *and the ideal of relations is generated by*

$$\{e_i(x_1,\ldots,x_n), 1 \le i \le n\}.$$

11.2.3 *The homology $H_*(Fl(n))$*

For $w \in W(= S_n)$, we shall denote by $[X_w]$ the class determined by the Schubert variety X_w in $H_{2l(w)}(Fl(n))$ (recall that $\dim X(w) = l(w)$). For $w \in W$, let Y_w be the closure of the B^--orbit $B^- \cdot e_w$, where B^- is the Borel subgroup opposite to B.

Theorem 11.2.3 (A geometric basis for $H_*(Fl(n))$). *The classes*

$$\{[X_u],\ l(u) = d\}$$

form a $\mathbb{Z}$-basis for $H_{2d}(Fl(n))$, and the classes $\{[Y_v],\ l(v) = d\}$ form a dual basis for $H_{2N-2d}(Fl(n))$ under the pairing

$$\langle\, ,\rangle : H_{2d}(Fl(n)) \times H_{2N-2d}(Fl(n)) \to H_0(Fl(n))(\cong \mathbb{Z}),$$

where $N = \binom{n}{2}(= \dim Fl(n))$.

11.2.4 *Schubert classes and Littlewood-Richardson coefficients*

Fix d, $1 \leq d \leq n-1$. Then the canonical projection $Fl(n) \to G_{d,n}$,

$$F_\bullet \mapsto V_d, \text{ where } F_\bullet = (V_0 \subset V_1 \subset \cdots \subset V_n = \mathbb{C}^n)$$

identifies $H^*(G_{d,n})$ as a subring of $H^*(Fl(n))$. Under this identification, for $\underline{i} \in I_{d,n}$, the Schubert class $[\sigma_{\underline{i}}]$ gets identified with the Schubert class $[X(w)]$ in

$$\mathbb{Z}[X_1, \ldots, X_n]/\langle e_i(X_1, \ldots, X_n), i = 1, \ldots, n\rangle,$$

where w is the permutation with the entries of $\underline{i}$ in the first d places, and the rest of the entries appearing in ascending order (such a permutation is called a *Grassmannian permutation*).

Let $G = SL_n(\mathbb{C})$. Recall (cf. Chapter 8) that the finite dimensional irreducible representations of $SL_n(\mathbb{C})$ are indexed by Young diagrams with columns of length at most $n-1$, two Young diagrams $\lambda = (\lambda_1 \geq \lambda_2 \geq \cdots \geq \lambda_{n-1} \geq 0)$, $\mu = (\mu_1 \geq \mu_2 \geq \cdots \geq \mu_{n-1} \geq 0)$ being identified if $\lambda_i - \mu_i$ is the same for all i. Let λ, μ be two Young diagrams. Let V^λ, V^μ denote the irreducible G-modules with highest weight λ, μ, respectively. Let

$$V^\lambda \otimes V^\mu = \bigoplus c_{\lambda\mu}^\nu V^\nu.$$

The integers $c_{\lambda\mu}^\nu$ are called the *Littlewood-Richardson coefficients*.

On the other hand, denoting the Schubert classes (in $H^*(Fl(n))$) corresponding to λ, μ, by $[\sigma_\lambda], [\sigma_\mu]$, respectively (recall that Schubert varieties in the Grassmannian correspond to Young diagrams, as in the discussion before Definition 7.5.10), we have that

$$[\sigma_\lambda] \cdot [\sigma_\mu] = \sum c^{\nu}_{\lambda\mu} [\sigma_\nu].$$

Thus the coefficients appearing in the product of two Schubert classes in $H^*(Fl(n))$ are simply the Littlewood-Richardson coefficients.

11.3 Free Resolutions

A classical problem in commutative algebra and algebraic geometry is to describe the syzygies of defining ideals of algebraic varieties. One typical example is the determinantal variety D_k of $m \times n$ matrices (over a field K) of rank at most $k(\leq min\{m,n\})$; it is a closed subvariety of the mn-dimensional affine space of all $m \times n$ matrices. When $K = \mathbb{C}$, a minimal free resolution of the coordinate ring $K[D_k]$ of D_k as a module over the coordinate ring of the mn-dimensional affine space (i.e., the mn-dimensional polynomial ring) was constructed by A. Lascoux [59]; see also [89, Chapter 6].

As seen in Chapter 10, a determinantal variety can be identified with the opposite cell in a certain Schubert variety in a suitable Grassmannian. In [42], a free resolution for a larger class of singularities, namely, *Schubert singularities*, i.e., the intersection of a singular Schubert variety and the "opposite big cell" inside a Grassmannian has been constructed. The method adopted in [42] is algebraic group-theoretic, and is likely to work for Schubert singularities in more general flag varieties. In fact, using the technique in [42], a free resolution for a class of Schubert singularities has been constructed in the Lagrangian Grassmannian in [33].

11.4 Bott–Samelson Scheme of *G*

Let G be a simple group of rank n. Denote the simple roots by $\{\alpha_1, \cdots, \alpha_n\}$, and the simple reflections by $\{s_1, \cdots, s_n\}$. For $1 \leq i \leq n$, let $P_i \supset B$ be the minimal parabolic subgroup of G associated to s_i.

For any word $\mathbf{i} = (i_1, \ldots, i_l)$, with letters $1 \leq i_j \leq n$, the *Bott–Samelson scheme* corresponding to the word $\mathbf{i}$ is defined as the quotient space

$$Z_{\mathbf{i}} = P_{i_1} \times P_{i_2} \times \cdots \times P_{i_l} \,/\, B^l \,,$$

where B^l acts by

$$(p_1, p_2, \ldots, p_l) \cdot (b_1, b_2, \ldots, b_l) = (p_1 b_1, b_1^{-1} p_2 b_2, \ldots, b_{l-1}^{-1} p_l b_l).$$

It was originally used (cf. [7, 18, 26]) to desingularize the Schubert variety $X_w = \overline{B \cdot wB} \subset G/B$, where $w = s_{i_1} \cdots s_{i_l}$. The desingularization is given by the multiplication map $m : Z_{\mathbf{i}} \to X_w \subseteq G/B$, $(p_1, \ldots, p_l) \mapsto p_1 \cdots p_l \cdot B$, and $Z_{\mathbf{i}}$ has the structure of an iterated fiber bundle with fiber $\mathbb{P}^1$ in each iteration, so we may loosely think of $Z_{\mathbf{i}}$ as a "factoring" of the Schubert variety into a twisted product of projective lines.

The Bott–Samelson scheme corresponding to the word $\mathbf{i} = (i_1, \ldots, i_N)$, $N = dim\, G/B$, where $s_{i_1} \cdots s_{i_N}$ is a reduced expression for the largest element of the Weyl group, is defined as the *Bott–Samelson scheme of* G. In recent times, many authors have studied Bott–Samelson varieties. In proving the Frobenius-split property (cf. §11.5 below) for *the generalized flag varieties* (namely, $G/B, G$ semisimple), the authors (cf. [73]) first prove the Frobenius-split property for Bott–Samelson varieties and then deduce the Frobenius-split property for the generalized flag varieties. In [79], the author proves the vanishing theorem for the cohomology of line bundles on Bott–Samelson varieties. A standard monomial theory has been developed for Bott–Samelson varieties in [47]; see also [2]. In [25], the authors study certain toric varieties associated to Bott–Samelson varieties.

In [27], the authors describe the Newton-Okounkov body of a Bott–Samelson variety as a lattice polytope defined by an explicit list of inequalities; a *Newton-Okounkov body* is a convex body in Euclidean space associated to a divisor (or more generally a linear system) on a variety. It encodes information about the geometry of the variety and the divisor. It is a generalization of the notion of the Newton polytope of a projective toric variety; for more details, see [36, 38, 60]. Besides Newton polytopes of toric varieties, several polytopes appearing in representation theory - such as Gelfand-Zetlin polytopes and string polytopes of Berenstein-Zelevinsky (cf. [3]), Littelmann (cf. [66]) - can be realized as special cases of Newton-Okounkov bodies. In fact, using the results of [3, 66], one gets a nice parametrization of Lusztig's canonical basis (cf. [69]), using which Caldero [11] describes toric degenerations of Schubert varieties; see also [15] for toric degenerations of Schubert varieties. For a related result (relating Okounkov bodies and toric varieties), see [1].

11.5 Frobenius-Splitting

Let X be a smooth projective variety over an algebraically closed field k of characteristic $p > 0$. Let $F : X \to X$ be the absolute Frobenius morphism induced by $f \mapsto f^p$ for all $f \in \mathcal{O}_X$. This induces an $\mathcal{O}_X$-linear map

$$F^{\#} : \mathcal{O}_X \to F_* \mathcal{O}_X.$$

An algebraic variety X is called a *Frobenius-split variety* if the $\mathcal{O}_X$-linear map $F^\#$ splits, i.e., if there exists an $\mathcal{O}_X$-linear map $s : F_*\mathcal{O}_X \to \mathcal{O}_X$ such that $s \circ F^\#$ is the identity map of $\mathcal{O}_X$.

The theory of Frobenius-splitting was developed by Mehta-Ramanathan (cf. [73]); it is a nice property implying many geometric and cohomological properties. Many varieties where a linear algebraic group acts with a dense orbit turn out to be Frobenius-split. This includes the generalized flag varieties and Bott–Samelson varieties; in fact, one first proves the Frobenius-split property for Bott–Samelson varieties and then deduces the Frobenius-split property for the generalized flag varieties by using the proper, birational map $m : Z \to G/B$ (cf. §11.4 above). Further, generalized flag varieties are split compatibly with their Schubert subvarieties (a subvariety Y of a Frobenius-split variety X is said to be *compatibly split in* X, if $s(F_*(I_Y)) \subset I_Y, I_Y$ being the ideal sheaf of Y (here s is as above the map $F_*\mathcal{O}_X \to \mathcal{O}_X$)).

The Frobenius-splitting of Schubert varieties yields important geometric results: Schubert varieties have rational singularities, and they are projectively normal and projectively Cohen–Macaulay in the projective embedding given by any ample line bundle (in particular, they are normal and Cohen–Macaulay). Further applications of Frobenius-splitting concern the representation theory of semisimple algebraic groups: for a detailed account, refer to [9].

11.6 Affine Schubert Varieties

Let $k = \mathbb{C}, F = k((t))$, the field of Laurent series, $A^\pm = k[[t^\pm]]$. Let G be a semisimple algebraic group over k, T a maximal torus in G, B a Borel subgroup, $B \supset T$ and let B^- be the Borel subgroup opposite to B. Let $\mathcal{G} = G(F)$. The natural inclusions

$$k \hookrightarrow A^\pm \hookrightarrow F$$

induce inclusions

$$G \hookrightarrow G(A^\pm) \hookrightarrow \mathcal{G}.$$

The natural projections

$$A^\pm \to k,\ t^\pm \mapsto 0$$

induce homomorphisms

$$\pi^\pm : G(A^\pm) \to G.$$

Let

$$\mathcal{B} = (\pi^+)^{-1}(B),\ \mathcal{B}^- = (\pi^-)^{-1}(B^-).$$

Let $\hat{W} = N(k[t, t^{-1}])/T$, the *affine Weyl group* of G (here, N is the normalizer of T in G); $\hat{W}$ is a Coxeter group on $\ell + 1$ generators $\{s_0, s_1 \cdots s_\ell\}$, where ℓ is the rank of G, and $\{s_0, s_1 \cdots s_\ell\}$ is the set of reflections with respect to the simple roots, $\{\alpha_0, \alpha_1 \cdots \alpha_\ell\}$, of $\mathcal{G}$.

We have the following *Bruhat decomposition*:

$$G(F) = \dot{\cup}_{w \in \hat{W}} \mathcal{B}w\mathcal{B},\ G(F)/\mathcal{B} = \dot{\cup}_{w \in \hat{W}} \mathcal{B}w\mathcal{B}(mod\, \mathcal{B}).$$

For $w \in \hat{W}$, let $X(w)$ be the *affine Schubert variety* in $G(F)/\mathcal{B}$:

$$X(w) = \dot{\cup}_{\tau \leq w} \mathcal{B}\tau\mathcal{B}(mod\, \mathcal{B}).$$

It is a projective variety of dimension $\ell(w)$.

11.7 Affine Flag and Affine Grassmannian Varieties

Let $G = SL(n)$, $\mathcal{G} = G(F)$, and $G_0 = G(A^+)$. Then $\mathcal{G}/\mathcal{B}$ is the *affine flag variety*, and $\mathcal{G}/G_0$ is the *affine Grassmannian*. Further,

$$\mathcal{G}/G_0 = \dot{\bigcup}_{w \in \hat{W}^{G_0}} \mathcal{B}w\, G_0(mod\, G_0)$$

where $\hat{W}^{G_0}$ is the set of minimal representatives in $\hat{W}$ of $\hat{W}/W_{G_0}$.

Denote $A^+ (= k[[t]])$ by just A. Let

$$\widehat{Gr(n)} = \{A\text{-lattices in } F^n\}$$

Here, by an A-lattice in F^n, we mean a free A-submodule of F^n of rank n. Let E be the standard lattice, namely, the A-span of the standard F-basis $\{e_1, \cdots, e_n\}$ for F^n. For $V \in \widehat{Gr(n)}$, define

$$\mathrm{vdim}(V) := \dim_k(V/V \cap E) - \dim_k(E/V \cap E).$$

One refers to $\mathrm{vdim}(V)$ as the *virtual dimension of* V. For $j \in \mathbb{Z}$ denote

$$\widehat{Gr_j(n)} = \{V \in \widehat{Gr(n)} \mid \mathrm{vdim}(V) = j\}.$$

Then $\widehat{Gr_j(n)}, j \in \mathbb{Z}$ give the connected components of $\widehat{Gr(n)}$. We have a transitive action of $GL_n(F)$ on $\widehat{Gr(n)}$ with $GL_n(A)$ as the stabilizer of the standard lattice E. Further, let $\mathcal{G}_0$ be the subgroup of $GL_n(F)$ defined as

$$\mathcal{G}_0 = \{g \in GL_n(F) \mid \mathrm{ord}(\det g) = 0\}$$

(here, for a $f \in F$, say $f = \sum a_i t^i$, ord(f) is the smallest r such that $a_r \neq 0$). Then $\mathcal{G}_0$ acts transitively on $\widehat{Gr_0(n)}$ with $GL_n(A)$ as the stabilizer of the standard lattice E. Also, we have a transitive action of $SL_n(F)$ on $\widehat{Gr_0(n)}$ with $SL_n(A)$ as the stabilizer of the standard lattice E. Thus we obtain the identifications:

$$(*) \qquad \begin{gathered} GL_n(F)/GL_n(A) \simeq \widehat{Gr(n)} \\ \mathcal{G}_0/GL_n(A) \simeq \widehat{Gr_0(n)}, SL_n(F)/SL_n(A) \simeq \widehat{Gr_0(n)}. \end{gathered}$$

In particular, we obtain

$$(**) \qquad \mathcal{G}_0/GL_n(A) \simeq SL_n(F)/SL_n(A)$$

Geometric and representation-theoretic results on affine Schubert, affine flag and affine Grassmannian varieties have been proved by several authors: [24, 39, 40, 61, 74].

References

1. Anderson, D.: Okounkov bodies and toric degenerations. Math. Ann. **356**(3), 1183–1202 (2013)
2. Balan, M.: Standard monomial theory for desingularized Richardson varieties in the flag variety $GL(n)/B$. Transform. Groups **18**(2), 329–359 (2013)
3. Berenstein, A., Zelevinsky, A.: Tensor product multiplicities, canonical bases and totally positive varieties. Invent. Math. **143**(1), 77–128 (2001)
4. Billey, S., Lakshmibai, V.: Singular Loci of Schubert Varieties. Progress in Mathematics, vol. 182. Birkhäuser, Boston (2000)
5. Billey, S., Warrington, G.: Maximal singular loci of Schubert varieties in $SL(n)/B$. Trans. Am. Math. Soc. **355**(10), 3915–3945 (2003)
6. Borel, A.: Linear Algebraic Groups. GTM, vol. 126, 2nd edn. Springer, New York (1991)
7. Bott, R., Samelson, H.: Applications of the theory of morse to symmetric spaces. Am. J. Math. **80**, 964–1029 (1958)
8. Bourbaki, N.: Groupes et Algèbres de Lie Chapitres 4, 5 et 6. Hermann, Paris (1968)
9. Brion, M., Kumar, S.: Frobenius Splitting Methods in Geometry and Representation Theory. Progress in Mathematics, vol. 231. Birkhäuser, Boston (2005)
10. Brown, J., Lakshmibai, V.: Arithmetically Gorenstein Schubert varieties in a minuscule G/P. Pure Appl. Math. Q. **8**(3), 559–587 (2012)
11. Caldero, P.: Toric degenerations of Schubert varieties. Transform. Groups **7**(1), 51–60 (2002)
12. Carrell, J.B., Kuttler, J.: Smooth points of T-stable varieties in G/B and the Peterson map. Invent. Math. **151**(2), 353–379 (2003)
13. Chevalley, C.: Classification de groups de Lie algébriques, Chevalley Séminaire 1956–58, vol. II. Secrétariat mathématique, Paris (1958)
14. Chevalley, C.: Sur les décompositions cellulaires des spaces G/B. Proc. Symp. Prue Math. **56**(Part I), 1–25 (1994)
15. Chirivi, R.: LS algebras and applications to Schubert varieties. Transform. groups **5**(3), 245–264 (2000)
16. De Concini, C., Eisenbud, D., Procesi, C.: Hodge algebras. Astérisque, vol. 91. Société Matématique de France, France (1982)
17. De Concini, C., Procesi, C.: A characteristic-free approach to invariant theory. Adv. Math. **21**, 330–354 (1976)
18. Demazure, M.: Désingularisation des variétés de Schubert généralisées. In: Annales scientifiques de l'École Normale Supérieure, vol. 7, pp. 53–88. Société mathématique de France, France (1974)

V. Lakshmibai, J. Brown, *The Grassmannian Variety*,
Developments in Mathematics 42, DOI 10.1007/978-1-4939-3082-1

19. Doubilet, P., Rota, G.C., Stein, J.: On the foundations of combinatorial theory. IX. Combinatorial methods in invariant theory. Stud. Appl. Math. **53**(3), 185–216 (1974)
20. Ehresmann, C.: Sur la topologie de certains espaces homogénes. Ann. Math. **35**, 396–443 (1934)
21. Eisenbud, D.: Commutative Algebra with a View Toward Algebraic Geometry. Graduate Texts in Mathematics, vol. 150. Springer, New York (2004)
22. Fulton, W.: Introduction to Toric Varieties. Annals of Mathematics Studies, vol. 131. Princeton University Press, Princeton (1993)
23. Fulton, W.: Young Tableaux: With Applications to Representation Theory and Geometry. London Mathematical Society Student Texts, vol. 35. Cambridge University Press, Cambridge (1997)
24. Gaussent, S., Littelmann, P.: LS galleries, the path model, and MV cycles. Duke Math. J. **127**(1), 35–88 (2005)
25. Grossberg, M., Karshon, Y.: Bott towers, complete integrability, and the extended character of representations. Duke Math. J. **76**(1), 23–58 (1994)
26. Hansen, H.C.: On cycles in flag manifolds. Math. Scand. **33**, 269–274 (1973)
27. Harada, M., Yang, J.J.: Newton-Okounkov bodies of Bott-Samelson varieties and Grossberg-Karshon twisted cubes. arXiv preprint. arXiv:1504.00982 (2015)
28. Hartshone, R.: Algebraic Geometry. Graduate Texts in Mathematics, vol. 52. Springer, New York (1997)
29. Herzog, J., Trung, N.: Gröbner bases and multiplicity of determinantal and Pfaffian ideals. Adv. Math. **96**(1), 1–37 (1992)
30. Hibi, T.: Distributive lattices, affine semigroup rings, and algebras with straightening laws. Commutative Algebra Comb. Adv. Stud. Pure Math. **11**, 93–109 (1987)
31. Hodge, W.: Some enumerative results in the theory of forms. Proc. Camb. Philos. Soc. **39**, 22–30 (1943)
32. Hodge, W., Pedoe, C.: Methods of Algebraic Geometry, vol. II. Cambridge University Press, Cambridge (1952)
33. Hodges, R., Laskhmibai, V.: Free resolutions of Schubert singularities in the Lagrangian Grassmannian. Preprint (2015)
34. Jantzen, J.: Representations of Algebraic Groups. Academic, New York (1987)
35. Kassel, C., Lascoux, A., Reutenauer, C.: The singular locus of a Schubert variety. J. Algebra **269**(1), 74–108 (2003)
36. Kaveh, K., Khovanskii, A.G.: Newton-Okounkov bodies, semigroups of integral points, graded algebras and intersection theory. Ann. Math. **176**(2), 925–978 (2012)
37. Kempf, G., Knudsen, F., Mumford, D., Saint-Donat, B.: Toroidal Embeddings. Lecture Notes in Mathematics, vol. 339. Springer, New York (1973)
38. Khovanskii, A.G.: Newton polyhedron, Hilbert polynomial, and sums of finite sets. Funct. Anal. Appl. **26**(4), 276–281 (1992)
39. Kreiman, V., Lakshmibai, V., Magyar, P., Weyman, J.: Standard bases for affine $SL(n)$-modules. Int. Math. Res. Not. **2005**(21), 1251–1276 (2005)
40. Kreiman, V., Lakshmibai, V., Magyar, P., Weyman, J.: On ideal generators for affine Schubert varieties. In: Algebraic Groups and Homogeneous Spaces. Tata Institute Fundamental Research Studies in Mathematics, pp. 353–388. Tata Institute Fundamental Research, Mumbai (2007)
41. Kumar, S.: The nil Hecke ring and singularity of schubert varieties. Invent. Math. **123**(3), 471–506 (1996)
42. Kummini, M., Lakshmibai, V., Sastry, P., Seshadri, C.: Free resolutions of some schubert singularities. arXiv preprint arXiv:1504.04415 [special volume of PJM dedicated to Robert Steinberg] (2015, to appear)
43. Lakshmibai, V.: Singular loci of Schubert varieties for classical groups. Bull. Am. Math. Soc. (N.S.) **16**(1), 83–90 (1987)
44. Lakshmibai, V.: Tangent spaces to Schubert varieties. Math. Res. Lett. **2**(4), 473–477 (1995)
45. Lakshmibai, V., Gonciulea, N.: Flag Varieties. Hermann, Paris (2001)

46. Lakshmibai, V., Littelmann, P., Magyar, P.: Standard monomial theory and applications. In: Proceedings of the NATO Advanced Study Institute on Representation theories and Algebraic Geometry, Series C: Mathematical and Physical Sciences, vol. 154, pp. 319–364. Kluwer Academic Publishers, Montreal (1997)
47. Lakshmibai, V., Littelmann, P., Magyar, P.: Standard monomial theory for Bott–Samelson varieties. Compos. Math. **130**(03), 293–318 (2002)
48. Lakshmibai, V., Magyar, P.: Degeneracy schemes, schubert varieties and quiver varieties. Int. Math. Res. Not. **12**, 627–640 (1998)
49. Lakshmibai, V., Musili, C., Seshadri, C.: Geometry of G/P–III. Proc. Indian Acad. Sci. **87A**, 93–177 (1978)
50. Lakshmibai, V., Musili, C., Seshadri, C.: Geometry of G/P–IV. Proc. Indian Acad. Sci. **88A**, 279–362 (1979)
51. Lakshmibai, V., Raghavan, K.: Invariant Theoretic Approach to Standard Monomial Theory. Encyclopedia of Mathematical Sciences, vol. 137. Springer, New York (2008)
52. Lakshmibai, V., Sandhya, B.: Criterion for smoothness of schubert varieties in $SL(n)/B$. Proc. Indian Acad. Sci (Math. Sci.) **100**, 45–52 (1990)
53. Lakshmibai, V., Seshadri, C.: Geometry of G/P–II. Proc. Indian Acad. Sci. **87A**, 1–54 (1978)
54. Lakshmibai, V., Seshadri, C.: Singular locus of a schubert variety. Bull. A.M.S. **11**, 363–366 (1984)
55. Lakshmibai, V., Seshadri, C.: Standard monomial theory and Schubert varieties, a survey. In: Proceedings of the CMS Conference on "Algebraic Groups and Applications" pp. 279–322. Manoj Prakashan (1991)
56. Lakshmibai, V., Seshadri, C.S.: Geometry of G/P. V. J. Algebra **100**(2), 462–557 (1986)
57. Lakshmibai, V., Weyman, J.: Multiplicities of points on a Schubert variety in a minuscule G/P. Adv. Math. **84**(2), 179–208 (1990)
58. Lang, S.: Algebra, revised 3rd edn. Addison Wesley, Massachusetts (1993)
59. Lascoux, A.: Syzygies des variétés déterminantales. Adv. Math. **30**(3), 202–237 (1978)
60. Lazarsfeld, R., Mustaţă, M.: Convex bodies associated to linear series. Ann. Sci. de lÉcole Norm. Supèrieure (4) **42**(5), 783–835 (2009)
61. Lenart, C., Postnikov, A.: Affine Weyl groups in K-theory and representation theory. Int. Math. Res. Not. IMRN **12** 65 (2007) (Art. ID rnm038)
62. Li, L., Yong, A.: Some degenerations of Kazhdan-Lusztig ideals and multiplicities of Schubert varieties. Adv. Math. **229**(1), 633–667 (2012)
63. Littelmann, P.: A generalization of the Littlewood-Richardson rule. J. Algebra **130**(2), 328–368 (1990)
64. Littelmann, P.: Good filtrations and decomposition rules for representations with standard monomial theory. J. Reine Angewandte Math. **433**, 161–180 (1992)
65. Littelmann, P.: A Littlewood Richardson rule for symmetrizable Kac-Moody algebras. Invent. Math. **116**(1–3), 329–346 (1994)
66. Littelmann, P.: Cones, crystals, and patterns. Transform. Groups **3**(2), 145–179 (1998)
67. Littelmann, P.: Contracting modules and standard monomial theory for symmetrizable Kac-Moody algebras. J. Am. Math. Soc. **11**(3), 551–567 (1998)
68. Littelmann, P.: The path model, the quantum Frobenius map and standard monomial theory. In: Carter, R., Saxl, J. (eds.) Algebraic Groups and Their Representations. Kluwer Academic Publishers, Dordrecht (1998)
69. Lusztig, G.: Canonical bases arising from quantized enveloping algebras. J. Am. Math. Soc. **3**(2), 447–498 (1990)
70. Manivel, L.: Le lieu singulier des variétés de Schubert. Int. Math. Res. Not. **2001**(16), 849–871 (2001)
71. Matijevic, J.: Three local conditions on a graded ring. Trans. Am. Math. Soc. **205**, 275–284 (1975)
72. Matsumura, H.: Commutative Ring Theory. Cambridge Studies in Advanced Mathematics, vol. 8. Cambridge University Press, Cambridge (1986)

73. Mehta, V.B., Ramanathan, A.: Frobenius splitting and cohomology vanishing for Schubert varieties. Ann. Math. (2) **122**(1), 27–40 (1985)
74. Mirkovic, I., Vilonen, K.: Perverse sheaves on affine Grassmannians and Langlands duality. arXiv preprint math.AG/9911050 (1999)
75. Mumford, D.: The Red Book of Varieties and Schemes. Lecture Notes in Mathematics, vol. 1358. Springer, New York (1999)
76. Mumford, D., Fogarty, J., Kirwan, F.: Geometric Invariant Theory, 3rd edn. Springer, New York (1994)
77. Musili, C.: Some properties of Schubert varieties. J. Indian Math. Soc. **38**, 131–145 (1974)
78. Newstead, P.: Introduction to Moduli Problems and Orbit Spaces, Tata Institute of Fundamental Research Lectures on Mathematics and Physics, vol. 51. Narosa Publishing House, New Delhi (1978)
79. Pasquier, B.: Vanishing theorem for the cohomology of line bundles on Bott-Samelson varieties. J. Algebra **323**(10), 2834–2847 (2010)
80. Perrin, N.: The Gorenstein locus of minuscule Schubert varieties. Adv. Math. **220**(2), 505–522 (2009)
81. Serre, J.: Faisceaux algebriques coherentses. Ann. Math. **61**, 197–278 (1955)
82. Seshadri, C.: Geometry of G/P–I, pp. 207–239. Springer, New York (1978) [Published by Tata Institute]
83. Spanier, E.H.: Algebraic Topology. McGraw-Hill Book Co., New York (1966)
84. Stanley, R.: Hilbert functions of graded algebras. Adv. Math. **28**, 57–83 (1978)
85. Stanley, R.: Enumerative Combinatorics. Cambridge Studies in Advanced Mathematics, vol. 2. Cambridge University Press, Cambridge (1999)
86. Sturmfels, B.: Gröbner Bases and Convex Polytopes. University Lecture Series, vol. 8. American Mathematical Society, Providence, RI (2002)
87. Svanes, T.: Coherent cohomology on Schubert subschemes of flag schemes and applications. Adv. Math. **14**, 269–453 (1974)
88. Weyl, H.: The Classical Groups. Their Invariants and Representations. Princeton University Press, Princeton (1939)
89. Weyman, J.: Cohomology of Vector Bundles and Syzygies. Cambridge Tracts in Mathematics, vol. 149. Cambridge University Press, Cambridge (2003)
90. Woo, A., Yong, A.: When is a Schubert variety Gorenstein? Adv. Math. **207**(1), 205–220 (2006)
91. Woo, A., Yong, A.: A Gröbner basis for Kazhdan-Lusztig ideals. Am. J. Math. **134**(4), 1089–1137 (2012)

List of Symbols

Symbols

V. Lakshmibai, J. Brown, *The Grassmannian Variety*,
Developments in Mathematics 42, DOI 10.1007/978-1-4939-3082-1

Index

V. Lakshmibai, J. Brown, *The Grassmannian Variety*,
Developments in Mathematics 42, DOI 10.1007/978-1-4939-3082-1

Printed in the United States
By Bookmasters